RANDOM
PROCESSES

STATISTICS: Textbooks and Monographs

A SERIES EDITED BY

D. B. OWEN, Coordinating Editor
Department of Statistics
Southern Methodist University
Dallas, Texas

PAUL D. MINTON
Virginia Commonwealth University
Richmond, Virginia

JOHN W. PRATT
Harvard University
Boston, Massachusetts

OTHER VOLUMES IN PREPARATION

RANDOM PROCESSES

A First Look

R. Syski

Department of Mathematics
University of Maryland
College Park, Maryland

MARCEL DEKKER, INC.   NEW YORK AND BASEL

Library of Congress Cataloging in Publication Data

Syski, R.   [Date]

    Random processes.

    (Statistics, textbooks and monographs; v.29)
    Includes index.
    1. Stochastic processes. I. Title.
QA274.S97   519.2   79-9129
ISBN 0-8247-6893-0

MARCEL DEKKER, INC.
270 Madison Avenue, New York, New York 10016

Current printing (last digit):
10 9 8 7 6 5 4 3 2

PRINTED IN THE UNITED STATES OF AMERICA

To inseparable brothers

Buc and Mieszko

who cherish random events

Many disciplines have their beginnings in the arcane strug-
gles of a small group of researchers. When the ideas generated
are seen to relate to human problems, they spread to a larger,
less specialized group. Then new questions develop, stimulating
both the growth of the discipline and its further spread.
Through such success and involvement, mathematics has penetrated
and shaped the physical and engineering sciences, growing itself
thereby. Even the softer sciences of biology, medicine, psychology, economics and management, traditionally qualitative, have
participated in this process, and today respond increasingly to
the demands of mathematics and its benefits.

In all these sciences, chance plays a key role, and the
special branch of mathematics devoted to its study, probability
theory, intrudes itself. One area of probability theory, sta-
tistical inference, deals largely with the implications of
randomness in fixed data for estimation and prediction. A second and somewhat newer area, stochastic processes, deals with
the dynamics of change in the presence of randomness. It stud-
ies entities fluctuating randomly in time such as temperature
levels, inventory levels, air indices, noise levels, delays in a
time-shared computer, etc. Such processes have been studied in
all the sciences mentioned above and are of growing importance.

Probability theory is one of the more difficult branches of mathematics in that its ideas are subtle and sometimes confusing. When this natural difficulty is aggrevated by a formal, abstract exhaustive exposition of the subject, it is placed beyond the reach of many of the people who need its tools. There has been a great need to present the basic ideas in a simple, lively manner, with the formal trappings of mathematics suppressed. As Professor Syski demonstrates, probability theory can be entertaining and does not have to be hidden in intellectual thickets.

The author is well qualified technically to write the kind of book needed. He is a highly competent mathematician and probabilist and has published extensively in the research literature. He has contributed to probabilistic potential theory, a very abstract and pure branch of probability theory. He has also contributed to the applied literature. His book on congestion theory was a pioneering work in this field, of value to both practitioners and theoreticians alike.

There are, however, many good probabilists, pure, applied and both. Two special human qualities are required to write such a book, vitality and humor. That Professor Syski has these qualities has long been known to his friends, and his astonishing enthusiasm for life, people and probability theory are very evident in this book.

J. Keilson
University of Rochester

It is common knowledge that observable phenomena in the
real life world -- from playing of games to the rise and fall
of empires -- are governed in their development to a large
extent by chance. The study of the underlying chance mechan-
isms lies in the domain of Probability Theory, and is called
the theory of random (or stochastic) processes. It is to the
credit of this theory that it can describe rather accurately
behavior of simple or complex systems, composed of living
organisms (human beings, animals, bacteria) or material objects
(machines, cars, equipment, etc.), subjected to chance fluctuations. These fluctuations have various origins both inter-
nal and external to the structure under consideration, and are
also related to the complexity of such a system.

Substantial progress has been made in recent years in
such areas as reliability (dealing with survival questions of
systems and their components), queueing (dealing with familiar
phenomena associated with waiting), traffic (telephone, road,
air, sea), inventories and storage (supply and demand fluctuations), medical aspects (spread of epidemics, survival and
treatment effectiveness), psychology (learning models),
economics and social sciences (human behavior).

The theory of stochastic processes owes its success to
solid mathematical foundations provided by modern probability

which itself made "a great leap" from gambling to respecta-
bility). Nowadays, the theory of random processes is a well-
established branch of mathematics with an extensively developed theoretical side (a paradise for abstract theoreticians)
and with impressive applications (for the practically minded
researchers). Numerous volumes have been written in this
field, ranging from elementary tests to sophisticated mono-
graphs, and a beginner must invest a great effort of time,
patience and determination to reach even the "first degree of
initiation."

There are, however, numerous people who for various reasons either cannot afford or do not wish to undergo such
demanding effort, but who would like to know for their own
enlightenment what the theory of stochastic processes is,
and who have (perhaps moderate) mathematical background sufficient to appreciate what is shown to them.

To these people this book is addressed. Its prerequis-
ite is an interest in the subject and a working knowledge of
elementary calculus. The book is indeed "the first look" at
random processes. It is not a systematic account of the
theory (such accounts can be found in existing textbooks),
but takes a light-hearted yet correct and moderately precise
approach to selected highlights of the theory. The book does
not offer a shortcut to the mastery of the subject (no such
shortcuts are possible!), but it presents only convincing
arguments (without rigors of formal proofs) in justification
of the announced results. Indeed, the book aims to develop
appreciation of the ingenuity involved in the mathematical

treatment of random phenomena, and of the power of the mathematical methods employed in the solution of applied problems. It is hoped that this book will stimulate readers' interest in further study of stochastic processes, by perhaps showing the applicability of methods and results to the readers' own field of interest. (Even if it fails in this respect, the casual reader who only glanced through its pages would benefit by gaining a new vocabulary.)

****

A large group of readers to whom this book is addressed are students not majoring in mathematics, but interested in applications of probability to their disciplines.

It may be desirable for non-math majors who have just completed a semester of calculus to see applications to the "real life" problems in social sciences, operations research and statistics. Such applications may have more appeal to some students than the routine examples from physics and engineering (usually found in calculus textbooks). This book has been designed to perform such a task with a two-fold aim:

i) to expose students to basic concepts of stochastic processes in an informal way through an intuitive approach with the help of calculus,

ii) to provide novel illustrative examples of applications of calculus to practical problems.

The primary object being applications, the book concen-

trates on the discussion of several selected problems. Selection has been made on individual merits, with problems lead-
ing to interesting mathematical techniques having first prior-
ity. However, topics extensively discussed in elementary
literature have been usually omitted; instead many problems
from advanced texts have been deliberately included.

Each problem is stated first in plain language. Then
its probabilistic model is developed, equations written down
and solved (or only the solution announced), and finally the
resulting formulae for probabilities or mean values are discussed and interpreted in practical terms. This essentially
is reduced to evaluation of integrals (integration by parts!),
differentiation (chain rule!), discussion of properties of
functions (slope, maxima and minima, inflection points, limits,
etc.), and solution of linear differential equations with
constant coefficients. From a probabilistic point of view,
readers get aquainted with such concepts as the distribution
function, density, random variables, expectation, moments,
etc.

No formal proofs are given, but only justification of
very loosely stated assertions. A justification is selected
if it presents an interesting example of calculus techniques;
otherwise, the result is shamelessly verified on one or two
examples and declared to be valid, especially if the final
result admits convincing practical interpretation. Numerical
calculations are strongly encouraged.

Although at first glance the problems may appear un-
related to each other, their common bond is supplied by

their probabilistic background. Indeed, the prevailing
theme is the distribution of various life times in renewal
theory, reliability and Markov chains.

   The material is arranged into four chapters, according
to thematic affinity. Chapter 1 (Easy Life and Good Times)
is devoted to properties of life times, and tacitly introduces
basic probabilistic concepts. Chapter 2 (Be Discreet with
Discrete) centers around Bernoulli trials and Poisson distribution, and their common applications. Chapter 3 (To Renew
or not to Renew) describes renewal theory and its ramifica-
tions. This chapter is more advanced than the preceding two,
and requires more attention from the reader. Chapter 4
(Markovian Dance) treats essentially examples of birth and
death processes with applications to queueing and learning
theory. Problems are treated separately and in order to
stress applications at this level, full power of Markovian
theory is regretfully not utilized. Each chapter begins with
a short introductory comment. The appendixes contain a list
of formulae and tables. On the whole, the book contains only
classical material, and therefore the individual references
are not given, but the interested reader should consult the
references listed in Appendix C (Suggestions for Further
Reading).

To the student

   This book aims to put some life in calculus techniques
you learned earlier, and which may appear to you rather dry.

It is also intended to show you that many real life situations can be analyzed with the help of probabilistic arguments using simple calculus.

Do not be horrified by strange looking formulae on the following pages. Many of them are already familiar to you, although may appear here in disguise. Others will become your friends soon, as you will study waiting times, queueing problems, learning, etc.

To the instructor

This is NOT a text in probability theory, nor is it a routine introduction to stochastic processes. This is simply a collection of several interesting problems which may appeal to imagination. Basic probabilistic concepts are introduced as they are needed, and in an unorthodox manner. Probability has been simply defined as the integral of density (area under a curve), or as a sum. A few new names have been invented to convey an idea (like "jumping rabbit" for a stopping time, "life times" for positive random variables, "cost function" for a function of a random variable.

If you have reservations against using such techniques like convolution integrals, double integrals, simple differential equations, etc., at this level, rest assured that they can be easily introduced in a very convincing manner. All that is needed is to tie them with the renewal process, combination of life times, birth and death processes, reliability, the Poisson process, and so on, in a very natural and simple way. Perhaps, using this simple interpretation of calculus concepts, the teaching of calculus may be more attractive.

It is surprising to see how such concepts can be appreciated by students at this level, and operated in a satisfactory manner, without heavy formal preparation.

Acknowledgments

The book is the outgrowth of a course on Stochastic
Models (Stat. 250) offered by the Statistics Branch, Department of Mathematics, University of Maryland, as a first introduction to applications of probability (with aims mentioned
in the Preface).

The author wishes to express his gratitude to his
colleagues in the Statistics Branch, associated with the
development of the Stat. 250 course, for their valuable comments. Special thanks are due to Karen Kirby who class tested
the present text in its early stages. The book also owes
much to the reaction of students in the Stat. 250 classes.

The author is grateful to Julian Keilson for his interest and his kind words; and shares with him his views on the
role of applied probability.

My warm thanks go to my son Marek who prepared the illustrations and drawings, and I am indebted to Pat Berg for her
superb typing of the manuscript.

R. Syski

Contents

Chapter 1

Easy Life and Good Times

In this chapter we shall discuss random fluctuations of
life times associated with people, animals and machines. As
we adopt a probabilistic point of view, it is appropriate to
begin with a precise formulation of our intuitive notion of
probability, and establish rules for calculations. Next, we
shall define what we mean by life time, and state how it can
be described in a convincing way. Then, we shall engage in
the study of numerous examples of diverse nature, involving
one life time, combinations of life times, and highly interesting associated features like cost functions, averages,
distribution functions, densities and other normal and abnor-
mal situations. Proceeding step-by-step, we shall develop
our probabilistic tools as the need arises. It is the
rewarding applications which will gratify our efforts in
acquiring these tools. So let us hope for easy life and
good times!

Perhaps we should stress again that in the probabilistic
description of the real life situations, we shall be rather
concerned with a mathematical model of the situation. The

real situation is too complex, and our tools rather crude,
so we must strike a balance between complexity of the situation and the complexity of our analysis. It should always
be remembered that our models are as good as the conclusions
which we can draw from them. If these conclusions are at
variance with observed facts, our model and our analysis are
of dubious validity.

Another fact should be stressed, too. We shall deal
with mathematical models and we shall need calculus. Para-
phrasing the inscription of the Platonian Academy, we can
say that there is no chance of studying probability without
calculus. So let us not hear complaints that what we are
doing is "just calculus and nothing but calculus!"

Section 1: Probability

Every person has some intuitive interpretation of the
meaning of the concept of probability (pr.). We may ask
such questions as, "what is the pr. that it will rain tomorrow," "what is the pr. that my favorite team will win a
particular game," "what is the pr. that I shall be killed
in a car accident," "what is the pr. that the waiting time
for a bus is less than 10 minutes," and so on.

When speaking about pr., the usual picture formed in
one's mind is that of dice throwing, card playing, or any
other games of chance. Although it is true that Probability
Theory originated from such considerations, today this theory is a well developed, self-contained mathematical disci-

pline with many applications to the physical sciences, engineering and the social sciences.

We shall be concerned here with a wide range of applications, ranging from the study of complicated systems of machines and people to the study of human behavior. We shall deal with topics which nowadays are generally called Operations Research. It is a far cry from dice and cards; as a matter of fact we shall ignore cards and dice almost entirely. Problems which we are going to discuss are important real life problems, studied extensively in the literature on probability.

Unfortunately, we must restrict ourselves to the simplest situations which can be handled with the limited mathematical tools at our disposal. We need rudimentary calculus -- this is a fact of life. Speaking about Probability without calculus is like taking a correspondence course in driving a car.

Intuitively, pr. is regarded as the opposition of certainty. We do not say that an event will take place (or took place), but we say only that such an event will (or did) probably occur. The pr. expresses our lack of information, our uncertainty, our degree of belief. Indeed, with all the evidence being the same, answers to the questions stated above may be entirely different, according to who is making the estimate of pr. Just think about the pr. of some controversial issue!

It will be necessary for us to eliminate from the start any reference to individuals who make estimates of pr. Thus,

the pr. of an event will be the same, irrespective of who
calculated it. This is done for purely technical reasons;
we shall, however, evaluate pr.'s of some events pertaining
to individuals, but we shall reformulate them appropriately.
In other words, we shall not discuss the subjective pr.,
but shall consider only the objective pr.

In comparing pr.'s of events, we may limit ourselves
to statements expressing the relative magnitude. Thus, we
may wish to know only that the pr. of one event is larger
than that of some other event. However, it is much more
natural and convenient, to express the pr. of an event by a
number. One can then operate with such numbers. For example, suppose that the pr. of waiting for a bus less than 10
minutes is 1/2, whereas for waiting less than one hour is
almost 1. How have these numbers been obtained? What do
they mean? How can one use them? That's what this book is
about.

The idea of attaching a number to a set or to an object
is familiar to everybody. We may talk about the length of a
segment of a line, the area of a triangle, the volume of a
sphere, or the mass of a physical body, its temperature, its
price, and so on. We may speak about the number of people
in the waiting room, the number of bacteria in a culture,
the time needed to memorize a particular poem, the number
of miles per gallon, the speed, the acceleration, the temperature, the pressure; the gain and the loss, the life and
death, pleasure and pain, the beauty and the beast -- all
these facts can be expressed in numbers.

Probability is also expressed in terms of numbers attached to events. The way of doing this is very much analogous to that of length, area and volume -- keep this in mind! Like these quantities, the pr. is nonnegative. If you have two plane figures which are disjoint, their total area is equal to the sum of their respective areas; if a farmer has 500 acres and his neighbor has 600 acres, their total holdings are 1100 acres. The volume of a sphere is equal to, say 15 cubic inches; any part of the ball necessarily has volume smaller than 15 cubic inches. It may be convenient to rescale the ball by saying that the volume is 1; then any part has volume expressed in fractions, or in percentage. Similarly, with the pr. One says that the pr. of an event which includes all events, the so called universal event or plainly the total event, is 1. Then, the pr. of any event is expressed by a fraction or a percentage. In other words, the pr. of a sure event (i.e., the event we know is bound to occur) is 1; the pr. of any other event lies therefore between 0 and 1. The exclusive events, that is events which cannot occur simultaneously are treated like disjoint sets; their probabilities add. Thus, if the pr. of having 5 people in the waiting room is 1/2 and the pr. of having 10 people is 1/3, the pr. of having either 5 or 10 people is 1/2 + 1/3 = 5/6. If the waiting room can hold 15 people only, the pr. of having 15 or less is 1. Note that the pr. of this room holding 20 people is 0, but of holding less than 20 is also 1.

In order to achieve a generality of discussion that is independent of the nature of events under consideration, it is convenient to denote events by letters (usually capitals) such as A, B, C etc. With our abbreviation of probability already introduced, we shall write the pr. of an event E in the form:

$$pr(E).$$

Thus, instead of the lengthy statement that "the waiting time for something is less than one hour with pr. 1/4," we shall write

$$pr(E) \ = \ 1/4$$

with the understanding that E represents the event in question. In general, for any event A

$$0 \ \leq \ pr(A) \ \leq \ 1.$$

Also, if events A and B are mutually exclusive, then

$$pr(A \ or \ B) \ = \ pr(A) \ + \ pr(B).$$

In particular, if A is an arbitrary event, then A^c will denote the complementary event; that is A^c means that A does not occur. For example, if A represents the event that there are exactly 6 machines in operation, the event

A^c means that the number of machines in operation is not 6. Since A and A^c are obviously mutually exclusive, one has the following useful relation:

$$pr(A) + pr(A^c) = 1.$$

How does one compute such numbers as pr(A)? There are certain rules and methods designed for this purpose and we shall talk about them as we solve several practical problems. Perhaps it may be of some interest to mention a rule which used to be of some importance in olden days, but has later been relegated to a remote place; yet it has some intuitive appeal and you may find it being used from time to time. Suppose that you wish to perform some simple experiment, like tossing a coin or a die, which may result in several outcomes. Suppose further that you are interested in a particular outcome only, say the head on a coin, or the 6 on a die. You perform this experiment many times and count the number of occurences of that particular outcome. Then, the pr. of the event A -- that is, the occurence of the outcome in question -- is a ratio of the number of occurences of that outcome to the total number of repetitions of the experiment. For example, if you toss a coin 100 times and the number of times a head appears is 57, then the pr. of a head is taken as 57%.

Such a procedure is rather vague, but supplies a pictureque intuitive background; it is known as the frequency

approach. It is very popular in applications, especially
in engineering. It fails miserably when the nature of
things prevents repetitions of the experiment. On the other
hand, under very special conditions (which we shall not
discuss here) one can show that when the number of repetitions of the experiment is sufficiently large, the frequency
approach will agree with the approach already mentioned
above. This is known as a law of large numbers.

For us, however, the pr. of an event is analogous to
length, area or volume; it is a number associated with an
event. In our discussions here, we shall investigate practical problems in which we shall be able to compute probabilities directly. Because of the highly involved structure of
real life situations, we shall restrict ourselves to investigations of rather simple situations. In other words, we
shall simplify our task by considering simplified versions
of the real life problems; that is, we shall study
probabilistic models. These models will be mathematically
simple, yet they will be sufficiently realistic.

We shall make some assumptions about the structure of
our model, and then derive several expressions for various
probabilities of interest. This will be the probability
part. As a result, we shall get a formula. To contrast it
with the real life situation, we shall use data collected
from observations, from experiments, etc., and shall draw
appropriate conclusions. This will be the statistics part.

Section 2:   Life Times

The presence of a time factor is a distinct feature of
many real life situations. When buying new equipment, say
a car, a radio or a dress, one usually asks how long it will
last. Sitting in the waiting room of a dentist, standing in
a queue in front of a ticket office, waiting for a bus --
all these are familiar situations involving waiting time.

More generally, we shall call a time needed to perform
some function, to observe something, a time for something
to happen -- a life time. This may be the actual life time
of equipment, a life time of a human being or of an animal;
or a waiting time for the arrival of a plane, a duration of
a telephone conversation, travel time between two cities,
and so on.

In all these situations there is an element of chance
involved. Indeed, the length of the life time fluctuates,
depending on many factors. Although one may be interested
in the exact duration of a life time, in most situations it
is enough to have some approximate estimate of this time.
You may be sceptical when told that a new gadget you just
bought will last exactly 30 days; you will be more convinced when told that it will probably last about a month.
Similarly, it is very likely that your new car will perform
well for about a year, and then repairs may ruin you; or
with bad luck somebody may hit your car the very moment you
leave the dealer's premises. Thus the exact length of time

your new car will operate until the first major breakdown
may be hard, or even impossible, to determine. You would
rather prefer to have some approximate estimate, say "about
a year;" or, checking published reports, you may find that
"on the average" that model of a car has such and such frequency of repairs.

Thus, it would be very desirable to have some way of
expressing quantitatively these possible fluctuations of
life times. It is here where Probability Theory enters.
Indeed, we shall talk about the pr. that a life time lies
within specified limits. For example, what is the pr. that
your new car will perform well for more than a month, for
more than a year? What is the pr. that within the length of
time you own this car, there will be one, two, ten major
repairs?

In other examples, one may ask for the pr. that the
waiting time for the first occurence of some event will be
larger than a specified amount? What is the pr. that a
telephone conversation will terminate within ten minutes
(when you are waiting outside the booth)?

In order to provide answers to such questions, it is
necessary to examine more closely the notion of a life time.
It will be convenient to write simply X for the life time.
Thus, X is a variable quantity whose fluctuations depend on
chance. Hence, our life time X is an example of a <u>random
variable</u>, a notion essential for Probability Theory.

We observe that the life time X is a positive quantity; it may be zero, of course. However, we shall ignore

a possibility of negative life times, although random variables assuming positive and negative values are very common
in other applications.

It is clear that a life time is finite, but we may
hesitate to impose any definite bound. A man can live 100
years, but not 200; what about 101, 102, ... ? 150? However, for convenience we shall assume that the life span is
finite, say of length L which may be one minute, ten years
or whatever is the case. Furthermore, we shall assume that
the life time X starts at some instant of time, "the origin," which will be denoted by s; say, today, tomorrow at
10 p.m., two years ago, and so on. Thus, our particular
life time X begins at the moment s and continues until
the instant s + L (when it terminates), its total duration
being L.

If the life time X is exactly of length t, where t
is a number between s and s + L, say t seconds, minutes,
years, etc., we shall write (X = t); here the symbol (X =
t) denotes the event that the life time X equals exactly
t. This type of symbolic notation is very economical. For
example, the event that the life time X has duration between a and b, inclusively, can be simply written in our
shorthand notation as:

$$(a \leq X \leq b).$$

Here a and b are numbers such that a is not larger
than b, and both are included between s and s + L. For
example, suppose that X represents a waiting time, being

measured from the instant zero until one hour; thus s = 0
and L = 60 (minutes). The event that one waits from 5
to 15 minutes is then written as (5 ≤ X ≤ 15). When
there is no possibility of confusion, we may even drop X
from our notation and denote the event in question simply
by [a,b]. In other words, we expressed our event concerning the duration of the waiting time, by an interval equal
to that duration. This is a very convenient procedure,
and we shall employ it frequently.

We now must associate pr. with the events concerning
X. In view of what has been said about probability in the
previous section, one must associate numbers with intervals
representing events. We are at liberty in choosing these
numbers in any reasonable way. At this stage, it is necessary to distinguish between two important modes of counting
time. We may measure time in discrete units, say 1, 2, 3
etc. (minutes, years, centuries), or continuously (in agreement with a poetic expression that "time flows"). These two
methods require different mathematical treatment.

We shall now restrict our discussion to the continuous
case. We shall express pr.'s of events with the help of a
function f, called the density of life time X. This
function f is defined on the interval [s,s+L] and has
the following two properties:

i) $f(t) \geq 0$ for every t such that $s \leq t \leq s + L.$

ii) $\int_{s}^{s+L} f(t)\ dt = 1.$

Property (i) means that f is never negative; property (ii) indicates that the area under the curve f(t) is equal to 1. Otherwise the function f is quite arbitrary, and we shall soon see several common examples.

With density f at our disposal, we shall define the pr. of the event that the life time X takes values between a and b, inclusively, as the integral of the density f taken from a to b; in symbols:

$$pr(a \le X \le b) = \int_a^b f(t)\, dt, \qquad where \quad s \le a \le b \le s+L.$$

Thus, the pr. of the interval [a,b] is just the area under the curve f(t) from a to b. In other words, we associated a number -- equal to the area under the curve -- to an interval.

Observe that by property (ii), the total area is 1 and this corresponds to the fact that the life time X must take some value between s and s + L, inclusively:

$$pr(s \le X \le s+L) = 1.$$

It follows from property (i) and properties of integrals that for intervals [a,b] ⊂ [a',b'], one would have

$$pr[a,b] \le pr[a',b'].$$

Similarly, if A and B are two disjoint intervals, then

$$pr(A \text{ or } B) = \int_A f(t)\, dt + \int_B f(t)\, dt \quad \text{the integration being}$$

taken over intervals A and B.

Of special interest are intervals of the form [s,t], where t varies from s to s + L, inclusively, and s is the initial point of the life time X. For this case we shall write:

$$pr(s \leq X \leq t) \quad = \quad \int_{s}^{t} f(x) \, dx, \qquad s \leq t \leq s + L.$$

Having fixed s and L, we may look at this integral as a function of t; denoted by F -- this function F is called the <u>distribution</u> <u>function</u> (d.f.) of a random variable (the life time) X. Thus,

$$F(t) \quad = \quad \int_{s}^{t} f(x) \, dx, \qquad s \leq t \leq s + L$$

and the number F(t) represents the pr. that the life time X is t or less; moreover, F(t) is equal to the area under the density curve taken from the origin s up to the point t.

Clearly,

$$0 \leq F(t) \leq 1, \quad \text{and} \quad F(s) = 0, \quad F(s+L) = 1.$$

Moreover, the d.f. F is always nondecreasing: $F(t) \leq F(t')$ whenever $t \leq t'$, because the interval [s,t] is contained in the interval [s,t'].

Observe also that

$$pr(a \leq X \leq b) \quad = \quad F(b) - F(a).$$

Of special interest is the event $(t < X \leq s+L)$, which is the complement of the event $(s \leq X \leq t)$. Indeed:

$$(s \leq X \leq t) \cup (t < X \leq s+L) = (s \leq X \leq s+L).$$

Hence, as we already noted in Section 1:

$$pr(t < X \leq s+L) = \int_t^{s+L} f(x)\, dx = 1 - F(t)$$

and this represents the area under the density curve from point t until the end. The expression $1 - F(t)$, denoted also by $F^c(t)$, is called the complementary d.f. in general, and the <u>survivor function</u>, with reference to the life time. Indeed, $F^c(t)$ gives the pr. that the life time has not failed up to time t. Clearly, $F^c(s) = 1$ and $F^c(s+L) = 0$; moreover, F^c is a nonincreasing function of t.

Observe that if the life time d.f. F is given, its density can be obtained by differentiation (for those t where f is continuous):

$$f(t) = dF(t)/dt$$

and similarly:

$$f(t) = -dF^c(t)/dt.$$

It should be stressed that $f(t)$ itself does not represent a pr.; indeed, the values of f may be larger than 1 for some t. However, the quantity $f(t)\, dt$ may be used to represent approximately the pr. that X assumes values between t and t + dt:

$$pr(t \leq X \leq t+dt) \; \approx \; f(t) \; dt;$$

the above expression is very useful in applications to practical problems.

Finally, observe that in the present case (when density f is given) the pr. that the life time X assumes exactly the value t is zero: indeed

$$pr(X = t) \; = \; \int_t^t f(x) \; dx \; = \; 0.$$

This is precisely the characterization of the continuous random variables X we now discuss.

It is now time to stop for a few examples.

Example 1: Uniform distribution.

$$f(t) \; = \; \begin{cases} \dfrac{1}{L} & \text{for} \quad s \leq t \leq s + L \\[2ex] 0 & \text{otherwise} \end{cases} \; ;$$

$$F(t) \; = \; \int_s^t \frac{1}{L} \, dx \; = \; \frac{t-s}{L} \qquad \text{for} \quad s \leq t \leq s+L;$$

$$F^c(t) \; = \; 1 - \frac{t-s}{L} \; ;$$

$$pr(a \leq X \leq b) \; = \; \int_a^b \frac{1}{L} \, dx \; = \; \frac{b-a}{L} \; .$$

Example 2: Negative exponential distribution (n.e.d.).
$(s = 0, L = \infty)$.

$$f(t) = \lambda e^{-\lambda t}, \quad t \geq 0, \quad \lambda > 0 \quad \text{constant};$$

$$F(t) = \int_0^t \lambda e^{-\lambda x}\, dx = 1 - e^{-\lambda t}, \quad t \geq 0;$$

$$F^c(t) = e^{-\lambda t}, \quad t \geq 0;$$

$$pr(a \leq X \leq b) = \int_a^b \lambda e^{-\lambda x}\, dx = e^{-\lambda a} - e^{-\lambda b}, \quad a \leq b.$$

Important simplification: It is a bit of a nuisance to remember the values of the initial instant s and of the duration L. In some cases we may take $s = 0$ (and we shall frequently do so), and we shall also encounter life times for which L is infinite. It is therefore mathematically more convenient to extend definitions of the d.f. F and its density f to the whole nonnegative real line, that is for all t from 0 to infinity. We shall simply write:

$$F(t) = 0 \quad \text{for} \quad 0 \leq t \leq s$$

and

$$F(t) = 1 \quad \text{for} \quad s+L \leq t < \infty$$

when L is finite. This is of course the same thing as defining

$$f(t) = 0 \quad \text{for} \quad 0 \leq t < s \quad \text{and for} \quad s+L < t < \infty;$$

indeed $f = 0$ does not contribute to the integral defining F.

Thus, we can define F by the integral:

$$F(t) = \int_0^t f(x)\ dx \qquad \text{for}\quad 0 \leq t < \infty$$

with

$$F(\infty) = \int_0^\infty f(x)\ dx = 1.$$

Clearly, this new definition preserves all properties of F mentioned above. Thus, $F(t)$ is a nondecreasing continuous function of t, and the value $F(t)$ represents the area under the density curve, up to point t. Moreover, this new definition of F has the advantage that it nicely takes care of life times for which L is infinite. Then, $F(t)$ is less than 1 for all real t, and approaches 1 as $t \to \infty$.

As before, we have now

$$F(b) - F(a) = \int_a^b f(x)\ dx$$

and

$$F^c(t) = \int_t^\infty f(x)\ dx.$$

The following graph shows a typical d.f. and its density.

Average life. The d.f. F of the life time X provides complete information about the behavior of X. However, it

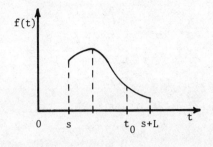

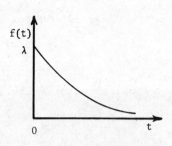

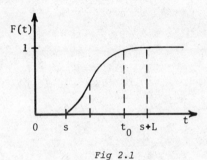

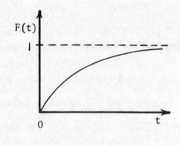

Fig 2.1

Fig 2.2

Typical distribution function F
and density f.

Exponential distribution.

is sufficient on many occasions to have only partial information about X. Thus instead of asking for the pr. of
waiting less than some value, it may be more informative to
know what the average waiting time would be.

We shall denote the <u>average</u> <u>life</u> <u>time</u> by μ or by
E(X) if dependence on X is stressed -- and define it by

$$\mu = \int_0^\infty tf(t) \, dt.$$

Here, one multiplies each value assumed by X with the
corresponding pr., say $t_n \cdot f(t_n) \cdot \Delta t_n$ for points t_n, and
sum over all n, in accordance with the notion of the
weighted average; the sum is replaced in the limit by the

integral. The words "the mean," the "expected life time"
are also used for μ. Observe that the defining integral
for μ is actually taken from s to s + L, when f
vanishes outside the interval [s,s+L].

Note that μ is a number associated with X. Different densities f may produce the same μ, so knowing μ
tells little about X. Conversely, if f is known, the
value of μ is of great interest in all applications of
life times.

In certain situations one may be interested in higher
moments μ_n defined by:

$$\mu_n = \int_0^\infty t^n f(t) \, dt$$

where $n = 0,1,2,\ldots$. Observe that always $\mu_0 = 1$, and
$\mu_1 = \mu$. The center of gravity, and the moment of inertia
known from physics, are analogues of moments in probability.
We shall almost never use moments of order higher than 2.

Since the life time X is a variable quantity, one may
be interested in its fluctuation around the mean. A convenient way of measuring this fluctuation is with the help of
a variance, denoted by σ^2, and defined as

$$\sigma^2 = \mu_2 - \mu^2 = \int_0^\infty (t-\mu)^2 f(t) \, dt.$$

Observe that the variance is never negative. It can be verified that the variance is small when most of the pr. is
concentrated around the mean; that is the graph of density

f has most of its area around the mean. Conversely, if the graph is spread around the mean, then the variance is large. (The positive square root of the variance, denoted by σ, is called the standard deviation.)

As an exercise in integration, one can obtain from the defining integral the equivalent expression for the mean (which are sometimes useful in computations)

$$\mu = \int_0^\infty F^c(t)\, dt \qquad \text{(integrate by parts)}$$

With reference to the examples mentioned earlier one finds by simple integration:

Example 1 -- continued:

$$\mu_n = \frac{1}{L}\int_s^{s+L} t^n\, dt = \frac{1}{L}\left.\frac{t^{n+1}}{n+1}\right|_s^{s+L} = \frac{(s+L)^{n+1} - s^{n+1}}{L(n+1)} \; ;$$

$$\mu = s + \frac{L}{2} \; ;$$

$$\sigma^2 = \frac{L^2}{12} \; .$$

Example 2 -- continued:

$$\mu_n = \int_0^\infty t^n \lambda e^{-\lambda t}\, dt = \frac{1}{\lambda^n}\int_0^\infty x^n e^{-x}\, dx = \frac{n!}{\lambda^n}$$

$$\mu = \frac{1}{\lambda} \, ,$$

$$\sigma^2 = \frac{1}{\lambda^2} \; .$$

Section 3:   Prolongation of Life Times

Suppose you wait in front of an occupied telephone
booth -- what is the pr. that the conversation which has been
in progress for some time, will end within the next couple
of minutes? You may ask what is the pr. that your car will
break down during the next month given that it performed
well since the last check up. If you treat a sick animal,
the natural question is to ask for the pr. of its survival,
when it is known that the animal is alive.

In these examples one actually considers the pr. that a
life time will be prolonged by an additional amount, given
that this life time has been continuing for some time. The
word "given" indicates that we calculate the pr. of some
event, given additional information -- in this case occurence
of another event. This is known as a conditional pr. If A
and B are two events, the event that A and B occur
simultaneously is denoted by A ∩ B. Its pr., written
pr(A ∩ B), could be taken as the pr. which interests us.
However, it is more convenient to work with the ratio of
pr(A ∩ B) to pr(B), where B is the conditioning event.
We thus define the conditional pr. of A given B by such
a ratio; in symbols, we shall write pr(A|B) -- which is
read "the pr. of A given B," the vertical line standing
for the word "given":

$$pr(A|B) \;=\; \frac{pr(A \cap B)}{pr(B)} \; .$$

(We must assume here that pr(B) is not equal to zero).

Thus pr(A|B) is the probability of the event A when it
is known that the event B has happened; in general this
pr. is different from pr(A) -- that event A happened (no
additional information).

Let us return now to our life time X. We shall again
represent our events concerning X by appropriate intervals.
As the origin of the life time will not matter much in our
consierations, we shall assume for convenience that the
initial point s = 0; this will be our standing convention
from now on. As in Section 2, we shall use F for the d.f.
and f for the density of the life time X. Thus X
assumes values from 0 to L, inclusive, where L is the
duration of X (which may be infinite); if L is finite,
then f(t) = 0 and F(t) = 1 for $t \geq L$.

Our task will be to find the pr. that the life time will
be prolonged by an additional amount, say h, when we know
that X lasted more than a specified value t. In terms of
events, (X > t+h) is the event that the life time is larger
than t + h; we shall take it as the event A in the above
definition. Similarly, (X > t) is the event that the life
time is larger than t; we take it for B. Thus, the pr.
we are going to find is:

$$pr(X > t+h \mid X > t).$$

In words, this is the pr. that the life time is continued
past t + h, given that it already lasted more than t. We
shall take this pr. as our defintiion of the pr. of prolongation by additional h; clearly $h \geq 0$ and $t \geq 0$.

Observe that $(X > t+h)$ implies $(X > t)$; indeed, if the life time lasted more than 15 years, it clearly lasted more than 10 years, but not conversely. Thus $(X > t+h) \cap (X > t) = (X > t+h)$. Substituting $A \cap B = (X > t+h)$ and $B = (X > t)$ in the above definition of the conditional pr.:

$$pr(X > t+h \mid X > t) = \frac{pr(X > t+h)}{pr(X > t)} .$$

Recall from Section 2 that in terms of the complementary d.f. F^C, these pr.'s are:

$$pr(X > t+h) = F^C(t+h), \qquad pr(X > t) = F^C(t).$$

Consequently, our pr. of prolongation by additional h is given by the formula:

$$pr(X > t+h \mid X > t) = \frac{F^C(t+h)}{F^C(t)} .$$

We shall now examine more closely this basic formula.

First of all observe that h can range from 0 to $L - t$, only (t is kept constant). For $h = 0$, the pr. of prolongation is 1, obviously; for $h = L - t$, the pr. of prolongation is evidently 0. Otherwise, the pr. decreases with increasing h -- this is also intuitive; indeed the longer prolongation of life time required, the smaller pr. of such an event. It should be noted that the pr. of prolongation depends in general on the instant t.

Examples: 1) For the case of uniform life (see Section 2):

$$f(t) = 1/L \quad \text{for} \quad 0 \le t \le L$$

and

$$F(t) = t/L \quad \text{for} \quad 0 \le t \le L.$$

Hence

$$F^C(t) = 1 - t/L \quad \text{and} \quad F^C(t+h) = 1 - (t+h)/L.$$

Consequently, the pr. of prolongation by h is:

$$F^C(t+h)/F^C(t) \;=\; \frac{L - t - h}{L - t} \;=\; 1 - h/(L-t)$$

for $0 \le h \le L-t$, and $0 \le t < L$.

2) For the case of exponential life (see Section 2):

$$f(t) \;=\; \lambda e^{-\lambda t} \quad \text{for} \quad 0 \le t$$

and

$$F(t) \;=\; 1 - e^{-\lambda t} \quad \text{for} \quad 0 \le t.$$

Hence

$$F^C(t) = e^{-\lambda t} \quad \text{and} \quad F^C(t+h) = e^{-\lambda(t+h)}.$$

Consequently, the pr. of prolongation by h is:

$$F^C(t+h)/F^C(t) \;=\; \frac{e^{-\lambda(t+h)}}{e^{-\lambda t}} \;=\; e^{-\lambda h}, \quad \text{for} \quad 0 \le h, \quad 0 \le t.$$

Observe that in this case the above pr. does not depend on t.   In other words, the pr. of prolongation by an additional amount does not depend on how long the life time has already lasted. For example, the pr. that the telephone conversation will terminate during the next 5 minutes does not depend on how long it has already been in progress. This perhaps need not be exactly true in real life situations, but there are other life times for which this property holds, at least approximately. Independence of the duration of a life time is the important property of the n.e.d.; it can be shown that the n.e.d. is the only distribution with this property. We express this property by saying that the exponential distribution is memoryless; it does not remember how long it lasted, as far as prolongation is concerned.

Return now to the general case. Suppose that we would like to know the pr. of a complementary event that the life time will last till t + h, given it lasts more than t. This is clearly equal to 1 - pr. of prolongation:

$$pr(X \le t+h \mid X > t) \; = \; 1 - F^C(t+h)/F^C(t) \; = \; \frac{F^C(t) - F^C(t+h)}{F^C(t)}$$

$$= \; \frac{F(t+h) - F(t)}{1 - F(t)}$$

for $0 \le h \le L - t$, and $0 \le t < L$.

Observe that this pr. equals 0 for h = 0, and equals 1 for h = L - t. Moreover, for fixed t, this pr. increases with h. Thus, as a function of h, it may be regarded as a <u>conditional</u> <u>d.f.</u> of a life time, given that

the life time lasted more than t. We shall denote this
conditional d.f. by K_t, the index t indicating dependence on t. Remember, however that t is kept fixed, and
it is h that varies! It is essential that t be strictly
less than L. Thus, recalling that F is the integral of
f, the above expression for $K_t(h)$ can be written as:

$$K_t(h) = \frac{1}{F^c(t)} \int_t^{t+h} f(x)\ dx, \quad \text{for } 0 \leq h \leq L-t, \quad 0 \leq t < L$$

with $K_t(0) = 0$, $K_t(L-t) = 1$ (for fixed t). Note that for
t = 0, one has $K_0(h) = F(h)$, and hence also $K_t(h) = 1$
for $h \geq L-t$.

Differentiation with respect to h (for fixed t)
yields the conditional density $k_t(h) = dK_t(h)/dh$, which
is given by

$$k_t(h) = \frac{f(t+h)}{F^c(t)}, \quad \text{for } 0 \leq h \leq L-t.$$

Here again, $k_t(h) = 0$ for $h > L-t$.

Of special interest is the value of this density at
h = 0, that is $k_t(0)$; it will be called the hazard rate
of the life time, and will be denoted by r(t). Thus

$$r(t) = \frac{f(t)}{F^c(t)}, \quad \text{for } 0 \leq t < L,$$

with r(t) = 0 for t > L. Indeed r(t) dh represents
approximately the pr. that a t-old life time will fail

immediately during dh;   r(t) is also called the <u>failure</u>
<u>rate</u>.

In problems concerning the reliability of equipment,
or components of equipment, one considers the <u>hazard</u> <u>func-</u>
<u>tion</u> R,   defined by

$$R(t) = \int_0^t r(x)\, dx, \quad \text{for } 0 \le t \le L.$$

The hazard rate r, as well as the hazard function R, determine uniquely the d.f. F and the density f of the
life time. It can be verified by a simple differentiation
that

$$F(t) = 1 - e^{-R(t)}, \quad \text{for } 0 \le t \le L.$$

Consequently, in agreement with above:

$$f(t) = r(t)e^{-R(t)}.$$

Thus, if the hazard rate r(t) is given, then the life time
d.f. can be easily computed. Note however, that F(L) = 1
implies that R(L) is infinite, so also R(t) = ∞ for
t ≥ L (for finite L).

Examples: 1) (Continuation) For the uniform life,

$$K_t(h) = h/(L-t), \quad \text{for } 0 \le h \le L-t$$

$$r(t) = 1/(L-t) \quad \text{and} \quad R(t) = -\log(1-t/L).$$

2)   (Continuation) For the exponential life,

$$K_t(h) = 1 - e^{-\lambda h}, \quad \text{for } 0 \leq h$$

and the hazard rate is constant

$$r(t) = \lambda$$

so $R(t) = \lambda t$, for $0 \leq t$.

Section 4: Bus Problem

Suppose that at a given stop buses are arriving independently of each other, the instants of their arrivals
distributed along a time axis. We can take the <u>inter-arrival</u>
<u>time</u> between two consecutive buses to be our life time X.
Indeed, the time interval between two consecutive buses
exhibits fluctuations which depend on chance, so X is a
random variable. We shall assume that all inter-arrival
times X have the same d.f. F with a density f and
(finite) mean μ; thus, μ is the average interval between
two consecutive buses.

The big question is now: what is the average waiting
time of a passenger arriving at some random instant at the
bus stop? The intuitive answer seems to be $\frac{1}{2} \mu$. Well,
this is NOT the case, except when buses arrive at exactly
fixed intervals of length μ. This is indeed a rather surprising result.

Once again: a passenger arrives at the bus stop
--naturally, there is no bus. The previous one has left a
long time ago and the next one will arrive (hopefully) soon.
The instant of the passenger's arrival is located at random
somewhere within the inter-arrival time (i.e. the life time)
X of buses. The passenger's <u>waiting time</u> W is of course
the time interval between his arrival and the arrival of the
next bus. Clearly, W is a random variable as its
fluctuations depend on chance. We can therefore consider
the event $(W \le t)$ that the waiting time is t or less and
we shall write for its probability:

$$G(t) = pr(W \le t).$$

In other words, the waiting time W is another life time,
and G is its distribution function. Our primary task will
be to find the form of G.

Despite its intuitive simplicity, the mathematical analysis of the problem is rather complicated, and cannot be
discussed here. Fortunately, the final result is surpris-
ingly simple, so with no hesitation we shall write down the
basic formula for G:

$$G(t) = \frac{1}{\mu} \int_0^t F^c(x) \, dx, \qquad \text{for } 0 \le t < \infty$$

Observe that G(t) increases with t, and that G(0) = 0,

$G(\infty) = 1$ as it should. Moreover, G has density
$g = dG/dt$ given by:

$$g(t) = F^c(t)/\mu, \quad \text{for } 0 \le t < \infty$$

with $g(0) = 1/\mu$.

In general, G differs from F. One should also avoid confusing G with pr. of prolongation discussed in Section 3; the situations are entirely different. Here we consider the remaining life time from a random instant till termination; thus W is sometimes called the residual life time.

We can now proceed to the evaluation of the average waiting time $E(W)$ which we simply denote by w. In accordance with what has been said about averaging (in Section 2), we must evaluate the integral

$$w = \int_0^\infty t\, g(t)\, dt.$$

Substituting for $g(t)$ and integrating by parts one has:

$$\int_0^\infty t\, g(t)\, dt = \frac{1}{\mu} \int_0^\infty t\, F^c(t)\, dt$$

$$= \frac{1}{\mu} \left[\frac{1}{2} t^2 F^c(t) \Big|_0^\infty + \frac{1}{2} \int_0^\infty t^2 f(t)\, dt \right]$$

$$= \frac{1}{2\mu} \int_0^\infty t^2 f(t)\, dt = \frac{1}{2\mu} \mu_2 ,$$

where μ_2 is the second moment of F

$$= \frac{1}{2\mu} (\sigma^2 + \mu^2),$$

because by definition $\sigma^2 = \mu_2 - \mu^2$. Consequently we have the final result:

$$\boxed{w = \frac{\mu}{2}(1 + \frac{\sigma^2}{\mu^2})}$$

Thus, w is always larger than $\frac{1}{2}\mu$, whenever σ^2 is not zero. The surprising fact is the dependence of the average waiting time w on the variance σ^2 of the life time distribution F. This in turn creates a paradox! For, if the d.f. F is such that $\sigma^2 > \mu^2$ (see Problem 5 for such an example), then clearly $w > \mu$, so the average waiting time is greater than the average inter-arrival time! It may even be infinite, if σ is infinite. Yet, as we defined it, the waiting time W is smaller than the inter-arrival time X.

Examples: 1) For uniform life time: As we have already seen:

$$F(t) = t/L, \quad \text{hence} \quad F^c(t) = 1 - t/L \quad \text{for} \quad 0 \le t \le L$$
$$= 1, \quad \text{hence} \quad = 0 \quad \text{for} \quad L \le t < \infty.$$

Moreover, $\mu = \frac{1}{2}L$, so

$$g(t) = (2/L)(1 - t/L) \quad \text{for} \quad 0 \le t \le L$$
$$= 0 \quad \text{for} \quad L \le t < \infty.$$

Furthermore, $\sigma^2 = L^2/12$, hence

$$w = L/3.$$

2) For exponential life: As we have already seen:

$$F(t) = 1 - e^{-\lambda t}, \quad \text{hence} \quad F^c(t) = e^{-\lambda t}, \quad \text{for} \quad 0 \le t < \infty.$$

Moreover, $\mu = 1/\lambda$, so

$$g(t) = \lambda e^{-\lambda t}, \quad \text{for} \quad 0 \le t < \infty.$$

Thus, we have another peculiarity of the n.e.d., namely that the densities f and g coincide. Indeed, the memoryless property of the exponential life implies that no matter when the passenger arrives, the distribution of the waiting time is exactly the same as if he would just miss the bus (and is forced to wait the whole inter-arrival time). The exponential life time does not pay any attention to late comers.

Since $\sigma^2 = 1/\lambda^2$, it is easy to check that $w = 1/\lambda = \mu$.

3) Constant life: This is a new example. Life times (inter-arrival times) are constant equal exactly to, say, L. No fluctuations are permitted (i.e. buses arrive at rigid fixed intervals. It is intuitively clear that now $\mu = L$, and as there are no fluctuations around the mean, the variance must be zero. Thus $\sigma^2 = 0$, hence $w = \frac{1}{2}\mu$. This is the intuitive answer mentioned at the beginning of this section.

But who ever saw buses arriving at regular intervals?

Section 5:   Combinations of Life Times

In many situations one must consider several life times simultaneously. If you have a device which is replaced when it breaks down, it may be of interest to know the total life span of several such replacements. In other situations, if several machines are in operation simultaneously, one may be interested in the shortest life time.

To handle such problems, information about joint behavior of life times is needed. This is provided by the joint density, or the joint d.f., of such life times. To be more specific, first consider two life times, say X and Y. The joint behavior of X and Y is described by their joint density f which is a function of two variables x and y. (For convenience, we shall used x and y as values of random variables X and Y, respectively). Clearly, $f(x,y) \geq 0$,   and

$$\int_0^\infty \int_0^\infty f(x,y) \ dx \ dy \ = \ 1.$$

We shall use our convention that f is 0 outside the range of values assumed by X and Y; restriction to nonnegative life times is nonessential.

Consider now the event, described jointly by X and Y, of the following form $(X \leq x, \ Y \leq y)$ -- this is the event that the first life time X is at most x, and the second life times Y is at most y. Write for the pr. of this event:

$$\mathrm{pr}(X \le x, \ Y \le y) \ = \ F(x,y)$$

and call F the joint d.f. of X and Y.

In agreement with our standard procedure of defining probabilities by assigning numbers to events, we shall define the above pr. F(x,y) by the double integral of the density f:

$$F(x,y) \ = \ \int_0^x \int_0^y f(t,s) \ dt \ ds, \qquad 0 \le x < \infty,$$
$$0 \le y < \infty.$$

Note that if F is given, f can be obtained by partial differentiation:

$$f(x,y) \ = \ \partial^2 F(x,y)/\partial x \partial y.$$

Suppose that the joint density of X and Y is given, then the density f_1 of X, as well as the density f_2 of Y, can be obtained from f by integration:

$$f_1(x) \ = \ \int_0^\infty f(x,y) \ dy, \qquad f_2(y) \ = \ \int_0^\infty f(x,y) \ dx.$$

It is customary to call f_1 and f_2 the marginal densities. Hence, clearly the (marginal) d.f.'s of X and Y, respectively, are:

$$F_1(x) = \int_0^x f_1(t)\, dt, \qquad F_2(y) = \int_0^y f_2(s)\, ds.$$

The joint d.f. F, or the density f, describes the
joint behavior of X and Y. From this joint behavior, as
the above formulas show, we can deduce the individual behavior of X and of Y, separately. It just suffices to find
the marginal densities (or d.f.'s). However, if we know
only the individual behavior of X and Y, separately,
nothing can be said in general about the joint behavior of X
and Y; in other words, knowledge of marginal densities
(d.f.'s) does not determine the joint density f (d.f. F).

There is, however, a very important special case when
marginals determine the joint distribution. This case is
referred to as <u>independence</u> of random variables X and Y,
and is characterized by the property that the joint density
f is simply the product of marginal densities f_1 and f_2:

$$f(x,y) = f_1(x) f_2(y), \qquad \text{for all } 0 \le x < \infty,$$
$$0 \le y < \infty.$$

This case introduces considerable mathematical simplifica-
tions. In the following we shall assume in most of our considerations that life times are independent. (We already
did this in the previous section with independent
inter-arrival times).

Simple (double) integration shows that independence
can be characterized alternatively in terms of d.f.'s by
factorization of the joint d.f. F:

$$F(x,y) = F_1(x)F_2(y), \quad \text{for all } 0 \le x < \infty, \quad 0 \le y < \infty.$$

Conversely, using this relation as a definition of independence, simple differentiation yields the product of densities, as stated earlier.

The following two remarks may be useful, although we shall not make much use of them now.

Remark 1: The function $f(x,y)$ describes a surface in the 3-dimensional space, and the integral $F(x,y)$ represents the volume under that surface up to the point (x,y). The total volume under the surface is taken to be 1.

Remark 2: (On independence). Two events A and B are said to be independent if the pr. of their joint occurrence is the product of their individual pr.s.

$$pr(A \cap B) = pr(A)pr(B).$$

Recall from Section 3, the definition of conditional pr.; independence implies that $pr(A|B) = pr(A)$, so A does not depend on B, in the probabilistic sense.

In agreement with our procedure of expressing events in terms of random variables, we can write:

$$A = (X \le x), \quad B = (Y \le y), \quad \text{so } A \cap B = (X \le x, Y \le y).$$

Hence, taking pr.s, one has from above:

$$pr(X \le x, \ Y \le y) \ = \ pr(X \le x) \cdot pr(Y \le y)$$

which is obviously

$$F(x,y) \ = \ F_1(x) F_2(y).$$

****

When the joint density is given, one can evaluate pr. of various events determined by life times X and Y, by simple integration. The general rule is as follows. Suppose that A is an event determined by X and Y. For example, $(X+Y \le z)$, or $(XY \le v)$, etc. These events A are in fact expressed by some function of X and Y, say $Z = \varphi(X,Y)$.

Next, determine the region in the x - y plane, corresponding to the event A; this again can be expressed in terms of the function $\varphi(x,y)$. Let $R(x,y)$ be this region. Then, the pr. of the event A is, in accordance with our procedure of assigning numbers to events, given by

$$pr(A) \ = \ \iint_{R(x,y)} f(x,y) \ dx \ dy$$

The integral being evaluated over the region $R(x,y)$. In general, calculations are rather involved, so we shall restrict ourselves to a few simple cases of interest for us. Our standing assumption is now that X and Y are independent.

****

Sum of two life times: Let X and Y be two independent life times, with densities f_1 and f_2, respectively. The total life time is $Z = X + Y$. It is required to find the density g of Z.

The answer is given by the following integrals, known as convolution of densities:

$$g(z) = \int_0^z f_1(x) f_2(z-x)\, dx = \int_0^z f_1(z-y) f_2(y)\, dy, \quad 0 \le z < \infty$$

This can be justified as follows. For the event $A = (X+Y \le z)$, the corresponding region is $R(x,y) = (x+y \le z)$, so according to the above procedure:

$$pr(X+Y \le z) = \iint_{x+y \le z} f_1(x) f_2(y)\, dx\, dy$$

$$= \int_0^z f_1(x)\, dx \int_0^{z-x} f_2(y)\, dy$$

using independence, and integrating first along the y-axis, then along the x-axis. The above integral gives, of course, the d.f. G(z) of Z. Hence, by differentiation $g(z) = dG/dz$ one obtains the first expression for g. The second follows in the same manner, by integrating first with respect to x, and then with respect to y. Note that here the region $(x+y \le z)$ is a triangle in the first quadrant, bounded by axes and the line $x + y = z$.

Computing the average total life time E(Z) it can be verified that

$$E(X+Y) = E(X) + E(Y).$$

So average life times add. (It can be shown that this property holds in general, irrespective of whether life times are independent.)

Note: For independent life times X and Y one has:

$$\text{var}(X+Y) = \text{var}(X) + \text{var}(Y).$$

This relation is NOT true in general for dependent life times.

****

In later sections we shall consider other combinations of life times. At the moment, let us stop for some examples.

****

Example 1: Suppose that X and Y are independent and have the same n.e.d.:

$$f_1(x) = \lambda e^{-\lambda x}, \quad 0 \leq x < \infty, \qquad f_2(y) = \lambda e^{-\lambda y}, \quad 0 \leq y < \infty.$$

Hence, the density of X + Y is:

$$g(z) = \int_0^z \lambda e^{-\lambda x}\, \lambda e^{-\lambda(z-x)}\ dx = e^{-\lambda z} \lambda^2 \int_0^z dx = (\lambda z) e^{-\lambda z} \lambda$$

for $0 \leq z < \infty$, and

$$E(Z) = \int_0^\infty z g(z)\ dz = \frac{2}{\lambda}.$$

Example 2: Suppose that X and Y are independent and have the same uniform density:

$$f_1(x) = \frac{1}{L}, \quad 0 \le x \le L \qquad\qquad f_2(y) = \frac{1}{L}, \quad 0 \le y \le L$$
$$\quad = 0, \quad x > L \qquad\qquad\qquad\qquad = 0, \quad y > L.$$

Hence, the density of X + Y is

$$g(z) = \int \frac{1}{L} \cdot \frac{1}{L} \, dx$$

where the integral is taken over the region such that simultaneously: $0 \le x \le L$, $0 \le z-x \le L$.

One must consider separately two regions (where always $0 \le z \le 2L$)

$$0 \le z \le L \qquad \text{for which} \qquad 0 \le x \le z$$
$$L \le z \le 2L \qquad \text{for which} \qquad z-L \le x \le L$$

so

$$g(z) = \int_0^z \frac{dx}{L^2} = \frac{z}{L^2} \qquad \text{for} \quad 0 \le z \le L$$

$$g(z) = \int_{z-L}^L \frac{dx}{L^2} = \frac{2L-z}{L^2} \qquad \text{for} \quad L \le z \le 2L$$

$$g(z) = 0 \qquad\qquad\qquad \text{for} \quad z > 2L.$$

The graph of g(z) is a triangle with vertices at points $(0,0)$, $(L, \frac{1}{L})$ and $(2L, 0)$.

Note: $E(X+Y) = 2 \cdot \frac{L}{2} = L.$

Section 6: Extreme Life Times

Suppose that in an experiment in Psychology there are
n subjects who must perform a certain task (say, children
working on a puzzle). The experiment is arranged so that
all n subjects start simultaneously, at instant 0 say,
and the objective is to find the shortest (as well as the
longest) time needed to perform the task. Of course, it is
not known at the beginning which of the subjects will finish
first, and which will finish last. It is therefore required
to find the distribution of the shortest time, and of the
longest time, needed (irrespective of the individual subjects).

Represent by $X_1, X_2, \ldots, X_n$ the length of time (life
time) needed to perform the task by subjects $1, 2, \ldots, n$,
respectively. Assume that life times $X_1, X_2, \ldots, X_n$ are
independent, and denote by $f_1, f_2, \ldots, f_n$ their densities,
and by $F_1, F_2, \ldots, F_n$ their d.f.'s.

The waiting time for the <u>first completion</u> is the smal-
lest life time out of $X_1, X_2, \ldots, X_n$. Denote it by M_-, so

$$M_- = \min(X_1, X_2, \ldots, X_n).$$

Clearly, M_- is a life time, and write G_- for its d.f.:

$$G_-(t) = \mathrm{pr}(M_- \le t), \qquad 0 \le t < \infty.$$

The event $(M_- > t)$ is the simultaneous realization of the n events $(X_1 > t), (X_2 > t), \ldots, (X_n > t)$ whose pr.'s are, respectively, $F_1^c(t), F_2^c(t), \ldots, F_n^c(t)$. Because of the assumed independence (see Section 5) the pr.'s multiply so

$$pr(M_- > t) = pr(X_1 > t) pr(X_2 > t) \ldots pr(X_n > t).$$

In other words, the complementary d.f. is:

$$G_-^c(t) = F_1^c(t) F_2^c(t) \ldots F_n^c(t)$$

and therefore $G_-(t) = 1 - G_-^c(t)$, for $0 \le t < \infty$. The corresponding density is found by differentiation: $g_-(t) = dG_-(t)/dt$.

In most cases of interest, the life times are identically distributed with the common d.f. F with density f. Consequently, the above expression simplifies to:

$$G_-(t) = 1 - [1 - F(t)]^n$$

and the corresponding density is

$$g_-(t) = n[1 - F(t)]^{n-1} f(t).$$

****

The waiting time for the <u>last completion</u> is the largest life time out of $X_1, X_2, \ldots, X_n$. Denote it by M_+, so

$$M_+ = \max(X_1, X_2, \ldots, X_n).$$

Clearly, M_+ is a life time, and write G_+ for its d.f.:

$$G_+(t) = pr(M_+ \leq t), \qquad 0 \leq t < \infty.$$

The event $(M_+ \leq t)$ is the simultaneous realization of the n events $(X_1 \leq t), (X_2 \leq t), \ldots, (X_n \leq t)$ whose pr.'s are, respectively, $F_1(t), F_2(t), \ldots, F_n(t)$. Because of the assumed independence (see Section 5) the pr.'s multiply so

$$pr(M_+ \leq t) = pr(X_1 \leq t) pr(X_2 \leq t) \ldots pr(X_n \leq t).$$

In other words, the d.f. is:

$$G_+(t) = F_1(t) F_2(t) \ldots F_n(t), \qquad 0 \leq t < \infty.$$

The corresponding density is found by differentiation $g_+(t) = dG_+(t)/dt$.

When the life times are identically distributed with the same F and f, then

$$G_+(t) = [F(t)]^n$$

and the corresponding density is:

$$g_+(t) = n[F(t)]^{n-1} f(t) \quad .$$

When distributions of extreme life times M_- and M_+ are known, one can compute their mean values $E(M_-)$ and

$E(M_+)$, as well as variances, using the method discussed in Section 2.

****

<u>Example</u>: Exponential life. All life times have the same n.e.d. (for $0 \le t < \infty$):

$$F(t) = 1 - e^{-\lambda t}$$

$$f(t) = \lambda e^{-\lambda t}$$

$$\mu = 1/\lambda.$$

Hence

$$g_-(t) = n\lambda e^{-n\lambda t}.$$

This is again the n.e.d., but with a parameter $n\lambda$, so

$$E(M_-) = 1/(n\lambda).$$

On the other hand

$$g_+(t) = n\lambda e^{-\lambda t}(1-e^{-\lambda t})^{n-1}, \qquad G_+(t) = (1-e^{-\lambda t})^n.$$

In order to compute the mean $E(M_+)$, it is more convenient to use the formula involving $G_+^c(t)$, as noted in Section 2 and in Section 4:

$$E(M_+) = \int_0^\infty G_+^c(t)\ dt = \int_0^\infty [1 - (1-e^{-\lambda t})^n]\ dt$$

$$= \int_0^1 (1-z^n)\ \frac{dz}{\lambda(1-z)},$$

after substitution of $\quad 1 - e^{-\lambda t} = z$

$$\lambda e^{-\lambda t} \, dt = dz$$

$$\lambda (1-z) \, dt = dz$$

$$= \frac{1}{\lambda} \int_0^1 (1+z+z^2+\ldots+z^{n-1}) \, dz,$$

because $\quad (1-z)(1+z+\ldots+z^{n-1}) = 1 - z^n$

$$= \frac{1}{\lambda} \left(1 + \frac{1}{2} + \frac{1}{3} + \ldots + \frac{1}{n}\right) \quad .$$

Note that for $\quad n \neq 1$:

$$E(M_-) < \frac{1}{\lambda} < E(M_+),$$

and that $\quad \lim_{n \to \infty} E(M_-) = 0, \quad \lim_{n \to \infty} E(M_+) = \infty.$

$$**************$$

The following practical problem is treated in exactly the same manner, stressing once more the fact that problems from different fields maybe mathematically identical.

Consider a system of n independent components, arranged either in series or in parallel (see figures) in which each component functions for a random length of time having a d.f. F, the same for all components, with density f. Thus, $F^c(t) = 1 - F(t)$ is the pr. that a component is functioning at time t (i.e., its life time is greater than t).

a) In series arrangement, the system functions when all its components are functioning.

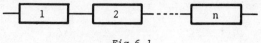

Fig 6.1

Series arrangement.

Thus, the d.f. G_s of the amount of time that the system functions is given by:

$$1 - G_s(t) = [1 - F(t)]^n.$$

b) In parallel arrangement, the system functions when <u>at</u> <u>least</u> <u>one</u> of its components is functioning.

Thus, the d.f. G_p of the amount of time that the system is functioning is given by:

$$1 - G_p(t) = 1 - [F(t)]^n.$$

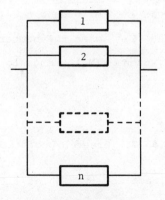

Fig 6.2

Parallel arrangement.

Note: expressions in (a) and (b) are known as <u>reliability</u>
of a system.

****

<u>Example</u>: Suppose now that life times of the components are
uniform: $F(t) = t/L$ (for $0 \leq t \leq L$). We have now for
$0 \leq t \leq L$:

(i) for series system:

$$1 - G_s(t) = (1 - \frac{t}{L})^n \, , \qquad g_s(t) = \frac{n}{L}(1 - \frac{t}{L})^{n-1} \, .$$

(ii) for the parallel system:

$$G_p(t) = (\frac{t}{L})^n \, , \qquad g_p(t) = \frac{n}{L}(\frac{t}{L})^{n-1} \, .$$

Denote by M_s and M_p the life time when the system functions, in series and in parallel arrangement, respectively.
These mean life times are given by

$$E(M_s) = \frac{L}{n+1} \, , \qquad E(M_p) = \frac{n}{n+1} L.$$

(Recall that the mean life of each component is $L/2$.)
Note that (for fixed L), when the number n of components
becomes very large, then $E(M_s)$ tends to zero, but $E(M_p)$
tends to L.

Section 7: Great Expectations

Frequently, direct observation of life time may be too
difficult to perform, or one may be more interested in some

other aspects of this life time. For example, recording
apparatus plots a graph of life time, subject to change of
scale; in design of a facility cost of life (waiting time)
may be of importance. Thus, the actual life time is
replaced by a secondary quantity which is some kind of a
function of the life time. For convenience, we shall call
it a cost function. Its values may be expressed in dollars,
or in some other units. In Economics, a term utility is
also used for the same purpose; we can also speak about gain
or loss, associated with life times. The term cost function
will embrace all such applications.

It may happen that in some situation, the pr. of a
long life time exceeding a fixed duration may be comfortably
large, but cost may be very large; to reduce costs, shorter
life times may be preferred. The most common method of
assessing costs is to evaluate the average cost of the life
time. We shall now discuss various aspects of such evalua-
tions.

Let X be a life time with density f, d.f. F and
mean life time μ. We shall denote a cost function by φ.
Thus, $\varphi(X)$ is a random variable representing cost associated with a life time X; in particular, if life time X
assumes the value x, then the cost function assumes the
value $\varphi(x)$. Observe that $\varphi(X)$ may assume values of both
signs (positive values representing gain, negative loss --
or conversely, depending on the point of view).

The average cost is now defined by:

$$E\varphi(X) \;=\; \int_0^\infty \varphi(x)\,f(x)\;dx\;.$$

Note that for $\varphi(x) = x$ the integral yields the usual mean μ, whereas for $\varphi(x) = x^n$ one obtains the moments μ_n (see Section 2). The point is that now we shall evaluate this integral for arbitrary functions φ.

Since $\varphi(X) = Y$ is a random variable, one could be interested in its distribution, and then evaluate $E(Y)$ as in Section 2; this of course can be done (using some calculus), but for evaluation of expectation it is, fortunately, unecessary. Perhaps it should be added that although the average cost has been defined above for a life time X, obviously the same definition applies to other life times, like the extreme life times (from Section 6), residual life times (from Section 4), and others, provided that f is now interpreted as an appropriate density.

Example 1: Linear Cost. Suppose that life is measured in days, and let a be the rate in dollars per day, and b the cost in dollars of initial operation. Thus, total cost for x days is (in dollars):

$$\varphi(x) = ax + b, \quad (x \geq 0).$$

The average cost is:

$$E\,\varphi(X) = \int_0^\infty (ax+b)\,f(x)\,dx = a\mu + b$$

irrespective of the form of density f.

Example 2: Quadratic cost. Suppose that m is a desirable

life of equipment in some construction project. Then, X - m

is a random fluctuation around m of the life time X. To

have positive values, it is convenient to consider rather

$(X-m)^2$ as the "error," departure from m. If c is the

unit cost, then the total cost of a discrepancy is $c(X-m)^2$.

The average cost is

$$Ec(X-m)^2 = c \int_0^\infty (x-m)^2 f(x) \ dx = c[\sigma^2 + (\mu-m)^2]$$

where $\sigma^2 = var(X)$. The integral is simply evaluated in the

following manner:

$$\int_0^\infty (x-m)^2 f(x) \ dx = \int_0^\infty [(x-\mu) + (\mu-m)]^2 f(x) \ dx$$

$$= \sigma^2 + 2(\mu-m) \int_0^\infty (x-\mu) f(x) \ dx + (\mu-m)^2$$

$$= \sigma^2 + (\mu-m)^2.$$

Suppose now that we wish to minimize the average cost by

selecting appropriate m. From the above expression for the

expected cost it is clearly seen that minimum will be

achieved when $m = \mu$. Thus, the mean life time minimizes

the quadratic cost, not matter what form of f.

Example 3: Kinetic energy of a particle of mass m is

given by $\frac{1}{2} mx^2$, where x is its velocity. Taking velocity as a life time   X,   the average kinetic energy is given by

$$E \frac{1}{2} m X^2 \;=\; \frac{1}{2} m \int_0^\infty x^2 f(x)\; dx \;=\; \frac{1}{2} m \mu_2$$

where μ_2 is the second moment.

Note that if X is the radius of a circle, the same argument gives for the average area $\pi \mu_2$ (which is less than $\pi \mu^2$, where μ is the average radius).

Example 4: Cut-off point. Suppose that a facility operates in such a way that if the waiting time X is less than a fixed amount a, then no loss is incurred. If the waiting time exceeds a, then the fixed penalty p is paid. The cost function is now:

$$\varphi(x) \;=\; 0 \qquad \text{for}\; x < a$$
$$=\; p \qquad \text{for}\; x \geq a.$$

The average cost is now:

$$E\varphi(X) \;=\; \int_0^a 0 f(x)\; dx + p \int_a^\infty f(x)\; dx \;=\; p F^c(a).$$

Suppose now that the penalty p is proportional to the cut-off point a, so p = ca, and that X has exponential life. Hence the average cost is

$$ca e^{-\lambda a},$$

and it is clear that the maximal average cost will occur at the cut-off point $a = \mu = 1/\lambda$, and equals $c\mu/e$.

Example 5: A signal produced by a device at time x is of the form $\sin \omega x$, where ω is the angular frequency. Hence, the average value of the signal for exponential life time X is:

$$E \sin \omega X = \int_0^\infty \lambda e^{-\lambda x} \sin \omega x \, dx = \frac{\lambda \omega}{\lambda^2 + \omega^2}$$

(the integral being evaluated by parts, twice).

Example 6: Strange life time. Suppose that density is given by the form:

$$f(x) = \frac{2}{\pi (1+x^2)} , \quad 0 \le x < \infty.$$

It is easy to verify that

$$\int_0^\infty f(x) \, dx = 1.$$

Let's look at its mean value:

$$\mu = \frac{2}{\pi} \int_0^\infty \frac{x}{1+x^2} \, dx = \frac{1}{\pi} \int_0^\infty \frac{dy}{1+y} = \frac{1}{\pi} \log(1+y) \Big|_0^\infty = \infty.$$

The mean is infinite! This may be a little puzzling, but intuitively it means that the average value is rather large, larger than any number. To cut it down to size, select a suitable cost function. Indeed, $\varphi(x) = \arctan x$ will do:

$$E(\arctan X) \;=\; \frac{2}{\pi} \int_0^\infty \arctan x \, \frac{dx}{1+x^2} \;=\; \frac{2}{\pi} \int_0^{\pi/2} y \, dy$$

$$=\; \frac{2}{\pi} \frac{1}{2} y^2 \Big|_0^{\pi/2} \;=\; \frac{\pi}{4} \,.$$

Example 7: One must be careful to avoid confusing $E(1/X)$ with $1/E(X)$. Suppose that X is exponential life, so

$$E(1/X) \;=\; \lambda \int_0^\infty \frac{1}{x} e^{-\lambda x} \, dx \;=\; \lambda \int_0^\infty \frac{1}{y} e^{-y} \, dy \;=\; \infty$$

(the integral is known to diverge). Similarly, for uniform life $f(x) = 1/L$ for $s \le x \le s+L$:

$$E(1/X) \;=\; \frac{1}{L} \int_0^{s+L} \frac{1}{x} \, dx \;=\; \frac{1}{L} \log x \Big|_s^{s+L} \;=\; \frac{1}{L} \log \frac{s+L}{s} \,.$$

and this is finite for $s > 0$, and infinite for $s = 0$.

Yet, in both cases $1/E(X) = 1/\mu$ which is finite.

Example 8: It is obvious that if the cost function is $\varphi(x) = x - \mu$, then the average cost is always zero, for any density f:

$$E(X-\mu) \;=\; \int_0^\infty (x-\mu) f(x) \, dx \;=\; \mu - \mu \;=\; 0.$$

****

When two life times X and Y are considered simultaneously, the cost function φ depends now on two variables, so $\varphi(X,Y)$. The average cost is then defined by the double integral:

$$E\varphi(X,Y) \quad = \quad \int_0^\infty \int_0^\infty \varphi(x,y)f(x,y) \ dx \ dy$$

where $f(x,y)$ is the joint density, as introduced in Section 5. Recall that in the important case when life times X and Y are independent, then $f(x,y) = f_1(x)f_2(y)$, where f_1 and f_2 are marginal densities of X and of Y, respectively (see Section 5).

Thus in the case of independence:

$$E\varphi(X,Y) \quad = \quad \int_0^\infty \int_0^\infty \varphi(x,y)f_1(x)f_2(y) \ dx \ dy.$$

Example 9: Linear cost. This is the extension of Example 1, with $\varphi(x,y) = ax + by + c$, where a and b are rates, and c is the initial cost. The average cost is:

$$E(aX+bY+c) \quad = \quad \int_0^\infty \int_0^\infty (ax+by+c)f(x,y) \ dx \ dy$$

$$= \quad \int_0^\infty axf_1(x) \ dy + \int_0^\infty byf_2(y) \ dy + c$$

$$= \quad aE(X) + bE(Y) + c$$

irrespective of the form of the joint density.

Example 10: Product cost. Suppose that cost is proportional to the product XY of life times, and that X and

Y are independent. Thus, $\varphi(X,Y) = cXY$, and the average cost is

$$E(cXY) = c \int_0^\infty \int_0^\infty xy f_1(x) f_2(y) \, dx \, dy$$

$$= c \int_0^\infty x f_1(x) \, dx \cdot \int_0^\infty y f_2(y) \, dy = cE(X)E(Y).$$

(This formula does not hold, in general, for dependent life times).

<u>Example 11</u>: Distance of a random point from the origin satisfies the relation $R^2 = X^2 + Y^2$, where (X,Y) are coordinates of a point in the first quadrant. The area of (a quarter) of a circle with radius R is $A = (\pi/4)R^2$. The average area is therefore

$$E(A) = \frac{1}{4}\pi \int_0^\infty \int_0^\infty (x^2+y^2) f(x,y) \, dx \, dy = \frac{1}{4}\pi [E(X^2) + E(Y^2)]$$

and this again depends on second moments (and not on square of the first moment); see Example 3.

<u>Example 12</u>: What is the average distance between two points chosen at random on the unit interval? To formulate this problem a little more precisely, suppose that the position of the points are represented by X and Y; the distance between them is $|X-Y|$, the absolute value is taken because distance is positive. Furthermore, we assume that X and Y are independent life times, each with a uniform

distribution on the interval from 0 to 1 (i.e., L = 1). Thus, the average distance is:

$$E|X - Y| = \int_0^1 \int_0^1 |x - y| \, dx \, dy = \frac{1}{3} .$$

Example 13: The analogous problem to Example 12 with n.e.d. on the positive half-axis leads to

$$E|X - Y| = \int_0^\infty \int_0^\infty |x - y| \, \lambda e^{-\lambda x} \, \lambda e^{-\lambda y} \, dx \, dy = \frac{1}{\lambda} .$$

The integral is rather tedious to evaluate, but the final result is of interest: the average distance is exactly equal to the average life.

Example 14: In contrast with Example 12 and Example 13, if cost is expressed as $(X - Y)^2$, the evaluation of average cost for independent X and Y is very easy:

$$E(X-Y)^2 = \int_0^\infty \int_0^\infty (x-y)^2 f_1(x) f_2(y) \, dx \, dy$$

$$= \operatorname{var}(X) + \operatorname{var}(Y) + (EX - EY)^2$$

irrespective of the form of densities.

Note on the Laplace transform

We have already seen that the exponential function has several special properties. It is therefore no surprise that the exponential cost function lives up to its name:

$$\varphi(x) = e^{-\alpha x}, \quad \text{for} \quad x \geq 0$$

where $\alpha \geq 0$ is a parameter. Physically, this cost function represents "declining costs" (or discount) at the rate α. The more important, however, is the average cost

$$E\varphi(X) = Ee^{-\alpha X}$$

of a non-negative life time X having density f. This average cost regarded as a function of α has rather remarkable properties, quite apart from its interpretation as a cost. Indeed, this new function, to be denoted by f^* and defined by:

$$f^*(\alpha) = \int_0^\infty e^{-\alpha x} f(x) \, dx \quad \text{for} \quad \alpha \geq 0$$

occurs frequently in applications, and deserves a special name; it is called the Laplace transform of the density f. Its primary usefulness is as a computational tool.

First of all note that no matter what form of the density f, its transform f^* is always a continuous function of α. Moreover, $f^*(0) = 1$.

Differentiation yields:

$$df^*(\alpha)/d\alpha = -\int_0^\infty x e^{-\alpha x} f(x) \, dx \ .$$

Hence, the value of the first derivative of f^* at $\alpha = 0$ yields the (negative of) the first moment of X:

$$df^*(\alpha)/d\alpha \Big|_{\alpha=0} = -E(X).$$

Similarly, the second derivative:

$$d^2f^*(\alpha)/d\alpha^2 = \int_0^\infty x^2 e^{-\alpha x} f(x) \ dx$$

yields the second moment:

$$d^2f^*(\alpha)/d\alpha^2 \Big|_{\alpha=0} = E(X^2).$$

More generally, the n-th differentiation yields:

$$d^n f^*(\alpha)/d\alpha^n \Big|_{\alpha=0} = (-1)^n E(X^n).$$

Thus, one can compute moments of X by differentiation -- and this is frequently very useful.

Using the expansion:

$$e^{-\alpha x} = \sum_{n=0}^\infty (-1)^n (\alpha x)^n / n!$$

it follows that, if all moments exist then:

$$f^*(\alpha) = \sum_{n=0}^\infty (-1)^n \alpha^n \mu_n / n!$$

so moments can be obtained from expansion of f^* in powers of α.

Furthermore, it is easy to check (by double integration) that if F is a d.f. with density f, then the Laplace transforms of F and F^c are, respectively:

$$\int_0^\infty e^{-\alpha x} F(x) \; dx \;\; = \;\; f^*(\alpha)/\alpha,$$

$$\int_0^\infty e^{-\alpha x} F^c(x) \; dx \;\; = \;\; [1-f^*(\alpha)]/\alpha \;.$$

Similarly, the Laplace transform of the derivative $f'(x) = df(x)/dx$ is

$$\int_0^\infty e^{-\alpha x} f'(x) \; dx \;\; = \;\; \alpha f^*(\alpha) \; - \; f(0).$$

Examples:

i) Uniform life:

$$f^*(\alpha) \;\; = \;\; \frac{1}{L} \int_s^{s+L} e^{-\alpha x} \; dx \;\; = \;\; \frac{e^{-\alpha s}(1-e^{-\alpha L})}{L\alpha} \;.$$

ii) Exponential life:

$$f^*(\alpha) \;\; = \;\; \int_0^\infty e^{-\alpha x} \lambda e^{-\lambda x} \; dx \;\; = \;\; \frac{\lambda}{\lambda+\alpha} \;.$$

iii) Constant life (see Section 4):

$$f^*(\alpha) \;\; = \;\; e^{-\alpha L}.$$

Laplace transforms are very useful in evaluation of properties of some cost functions or combination of life times, provided of course that resulting expressions are easier to work with than those obtained by direct methods.

For example, if Y is a linear transformation of a life time X, that is $Y = aX + b$, then the transform of the density of Y can be expressed in terms of the transform of density of X as follows:

$$Ee^{-\alpha Y} \;=\; Ee^{-\alpha(aX+b)} \;=\; e^{-\alpha b}Ee^{-\alpha aX} \;=\; e^{-\alpha b}f^{*}(\alpha a).$$

From this moments of Y can be easily computed.

The following expressions (stated here without proof) are very useful:

$$\lim_{\alpha\to\infty} f^{*}(\alpha) \;=\; 0, \qquad \lim_{\alpha\to\infty} \alpha f^{*}(\alpha) \;=\; \lim_{t\to 0} f(t),$$

$$\lim_{\alpha\to 0} \alpha f^{*}(\alpha) \;=\; \lim_{t\to\infty} f(t).$$

Another illustration of advantages offered by Laplace transforms is provided by the transform of a convolution integral. Recall from Section 5 that if X and Y are two independent life times with densities f_1 and f_2 respectively, then the total life time $Z = X + Y$ has a density g given by convolution integral

$$g(z) \;=\; \int_{0}^{z} f_1(x)f_2(z-x)\,dx.$$

In the present case, the Laplace transform of g is given simply by the product of the Laplace transforms of f_1 and of f_2, that is:

$$g^*(\alpha) \;=\; f_1^*(\alpha) f_2^*(\alpha).$$

This is easily justified as in Example 10 above, making use of independence:

$$Ee^{-\alpha Z} \;=\; Ee^{-\alpha(X+Y)} \;=\; E(e^{-\alpha X}e^{-\alpha Y}) \;=\; Ee^{-\alpha X}\cdot Ee^{-\alpha Y}.$$

For those who like double integrals, the same result may be obtained alternatively by direct evaluation of the integral, by using the explicit form of density g. Thus:

$$g^*(\alpha) \;=\; \int_0^\infty e^{-\alpha z}g(z)\ dz \;=\; \int_0^\infty e^{-\alpha z}\ dz \int_0^z f_1(x)f_2(z-x)\ dx$$

$$\;=\; \int_0^\infty f_1(x)e^{-\alpha x}\ dx \int_z^\infty e^{-\alpha(z-x)}f_2(z-x)\ dz$$

$$\;=\; f_1^*(\alpha) f_2^*(\alpha).$$

As an example, consider Example 1 in Section 5. The Laplace transform is simply

$$g^*(\alpha) \;=\; \left(\frac{\lambda}{\lambda+\alpha}\right)^2 .$$

****

For further applications of Laplace transforms see
Sections 17 and 24.

Section 8: Double Scotch

Recall that in Section 5 we have defined the pr. of an
event determined by two random variables X and Y as an
integral of their joint density f over a set R in the
plane where X and Y take their values:

$$\text{pr}(A) \;=\; \iint_{R(x,y)} f(x,y)\;dx\;dy.$$

In other words, with each set R we associate a real number, the value of the above integral as the pr. of the
corresponding event A.

The integral to be evaluated is a double integral, and
in this section some illustrative examples will be worked
out.

1. <u>Two friends meeting</u>. Suppose that two friends on our
Campus want to meet in front of the Math building at noon-
time. Class locations and traffic being what they are, the
friends can only count on arriving sometime between noon
and 1 p.m. They agree to show up at the steps of the
building somewhere in that interval, with the stipulation
that, in order not to waste too much time waiting, each
will wait only 10 minutes after arriving and then leave
if the other has not shown up. What is the pr. of their
actually meeting?

To make this problem precise, we shall assume that each of the two friends may arrive any time within the hour from noon to 1 p.m., all times of arrival being distributed with the same uniform density (with $L = 1$ hour). Thus, we have here life times (arrival times) of two friends, to be denoted by X and Y, and each of these random variables has the uniform distribution.

The second assumption to be imposed is that the arrival times of the two friends are completely independent of each other. This assumption that X and Y are independent life times is of crucial importance; it is expressed by the relation $f(x,y) = f_1(x)f_2(y)$. In the present case the joint density is

$$f(x,y) = 1/L^2, \qquad 0 \le x \le L, \qquad 0 \le y \le L$$

(and $f = 0$ outside the square with sides L).

Now, $X - Y$ is the distance between the friends arrivals. Since our problem requires that distance between arrivals be positive (and not exceed 10 min.) we must take the absolute value $|X-Y|$. The event A we look for is $|X-Y| \le h$, and its pr. is according to the above rule:

$$\Pr(|X-Y| \le h) \;=\; \iint_{|x-y| \le h} (1/L^2) \; dx \; dy$$

the region $R(x,y)$ in the plane being determined by inequalities:

$$|x-y| \le h, \qquad 0 \le x \le L, \qquad 0 \le y \le L.$$

Note that as a function of h, the above integral gives
the d.f. of life time |X-Y|.

In order to evaluate this double integral, one could
proceed formally, but in the present case a simple observation dispenses with any integration. First note that $1/L^2$
is a constant, so may be taken out from the integration
sign; the integral then gives simply the area of the strip
R. Next, notice that the strip R is obtained from the
square of sides L, by removing two equal triangles of
equal sides of length L - h. The area of a tringle being
well known, we have:

$$\text{pr}(|X-Y| \leq h) = (1/L^2)(L^2 - (L-h)^2) = 1 - (1-h/L)^2$$

for $0 \leq h \leq L$.

In our problem, L = 1 hour, h = 10 min. = 1/6 hour,
so from the above formula, the required pr. is 11/36.

Observe also that in the present case density is
(1-h/L) 2/L, and the mean waiting time has been already
calculated in Example 12 in Section 7, and equals L/3 =
20 minutes.

2. <u>Quadratic equation</u>. We know from algebra that a quadratic equation

$$Ax^2 + Bx + Cx = 0$$

can have two, or one or none real roots. Suppose you write

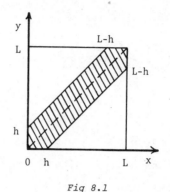

Fig 8.1

The event A *that meeting takes place (shaded area).*

down an equation choosing its coefficients at random. What
is the pr. that it has two real roots? Of course, you can
immediately check if the given equation has two real roots,
and no probability is needed. The problem is, however, more
general -- we actually ask how frequently you may expect an
equation to have two, one or no real roots. (The question
is far from being trivial, and for equations of higher
degree only approximate answers are known.)

We shall solve this problem in a very simple case.
First of all observe that coefficients A, B, and C are
random variables. Our first simplification will be to take
A = 1, a constant. Thus our equation can be rewritten as

$$x^2 + Bx + C = 0$$

where B and C are random variables. We need their joint
density. Here comes our second, rather restrictive assumption. Namely, we shall take B and C as independent,

each with uniform density over the unit interval [0,1].
Thus, the joint density is simply 1, over the unit square.

The quadratic equation has two real roots when
$B^2 - 4C > 0$; this is therefore the event whose pr. we must
evaluate. It is simply represented as the double integral
over the region R determined by the inequality

$$b^2 - 4c > 0, \qquad 0 \le b \le 1, \qquad 0 \le c \le 1$$

within the unit square. Hence:

$$\mathrm{pr}(B^2 - 4C > 0) = \iint_{b^2 - 4c > 0} db\ dc .$$

The indicated region is bounded by the parabola $c = \frac{1}{4} b^2$
and the b-axis, and the above integral is simply the area
of this region. Hence, its value is:

$$\int_0^1 \int_0^{\frac{1}{4}b^2} dc\ db = \frac{1}{4} \int_0^1 b^2\ db$$

$$= \frac{1}{12} .$$

Hence the answer to our problem is 1/12.

Note that $B^2 - 4C = 0$ is the condition to have
exactly one real root; the same argument gives the
integral over the parabola $b^2 - 4c = 0$, and its
value is of course 0. Consequently, it is unlikely to
pick up an equation with only one root. On the other hand,

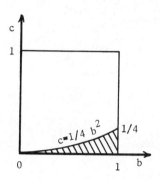

Fig 8.2
The region $b^2 - 4c > 0$ (shaded area).

the pr. of an equation having no real roots is obviously 11/12.

Clearly, this answer will change with a different choice of distribution of coefficients.

3. <u>Double trouble</u>. When comparing two life times X and Y it is of importance to know pr.'s of events like X > Y or X = Y. These pr.'s are given by:

$$pr(X > Y) \;=\; pr(X-Y > 0) \;=\; \iint_{x-y>0} f(x,y) \; dx \; dy$$

$$pr(X=Y) \;=\; \iint_{x=y} f(x,y) \; dx \; dy \;=\; 0.$$

In the case of independence, the first integral reduces to

$$\int_0^\infty f_1(x)F_2(x) \; dx = 1 - \int_0^\infty f_2(y)F_1(y) \; dy. \quad \text{Suppose now that}$$

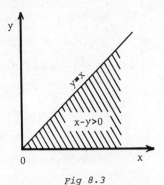

Fig 8.3

The event X > Y (shaded area).

X and Y are independent with the same density f. Then
(with z = F(x)):

$$\Pr(X > Y) \;=\; \int_0^\infty f(x)\,F(x)\;dx$$

$$=\; \int_0^1 z\;dz \;=\; \tfrac{1}{2}.$$

where z = F(x), dz = f(x) dx, $0 \le x < \infty$, $0 \le z \le 1$.
It is of interest that this pr. is always $\tfrac{1}{2}$ for any density
f.

Section 9: How Normal Is Normal?

So far we have dealt exclusively with life times which
are nonnegative. Many situations require, however, introduction of negative values. For example, the difference of
two positive life times may assume negative values; also if
the time origin is within a life time, negative values

represent simply the time before the origin. In this Section a few comments on life times of arbitrary sign will be
made. It is more convenient to speak in such a situation
about a <u>random variable</u> (r.v.) rather than about life time;
see Section 2. If X is a r.v., its density f is now
defined for all x, from $-\infty$ to $+\infty$, and its d.f. F
now has this form:

$$F(x) = \int_{-\infty}^{x} f(t)\ dt, \quad -\infty < x < \infty$$

and the moments are defined by

$$\mu_n = \int_{-\infty}^{\infty} x^n f(x)\ dx.$$

Similarly, the average cost is now:

$$E\,\varphi(X) = \int_{-\infty}^{\infty} \varphi(x)\,f(x)\ dx.$$

This is actually where similarities end; other formulae
which we discussed so far refer essentially to nonnegative
life times.

****

We shall now restrict our discussion to the famous
Normal (or Gaussian) distribution, of great importance in
Statistics and Probability.

The normal density is defined for all real x by a
single formula:

$$f(x) = \frac{1}{\sqrt{2\pi}\,\sigma}\, e^{-\frac{(x-\mu)^2}{2\sigma^2}}, \quad -\infty < x < \infty,$$

where μ and σ are two constants such that: $-\infty < \mu < \infty$ and $0 < \sigma < \infty$.

The graph of f is the familiar bell-shaped curve, symmetric with respect to the vertical line $x = \mu$. Indeed, $f(x) = f(2\mu - x)$. Furthermore, simple differentiation yields:

$$f'(x) = -\frac{x-\mu}{\sigma^2}\, f(x), \qquad f''(x) = \frac{1}{\sigma^4}\,[(x-\mu)^2 - \sigma^2]\, f(x).$$

It follows that the curve $f(x)$ has maximum at $x = \mu$, equal to $f(\mu) = \dfrac{1}{\sqrt{2\pi}\,\sigma}$, and two inflection points at $\mu + \sigma$, and $\mu - \sigma$. Moreover, the width of the bell-shaped curve is proportional to the parameter σ; indeed, the distance between two inflection points is 2σ.

There is no nice formula for the d.f. F; one must write out the full integral. It is convenient to introduce the so called <u>standard</u> form of the normal density, denoted by ϕ, which is obtained by taking $\mu = 0$ and $\sigma = 1$:

$$\phi(x) = \frac{1}{\sqrt{2\pi}}\, e^{-\frac{x^2}{2}}, \quad -\infty < x < \infty.$$

The corresponding standard form of the d.f. is of course

$$\Phi(x) = \int_{-\infty}^{x} \phi(t)\, dt.$$

Returning now to the arbitrary μ and σ, one finds that

the d.f. F can be expressed in terms of a standard d.f.
by the relation:

$$F(x) \;=\; \Phi((x-\mu)/\sigma) \;.$$

This follows by the change of variable $(t-\mu)/\sigma = y$.

The values of the standard distribution are tabulated
so using tables one can compute $F(x)$ for any μ and σ.
One can check that $F(\mu+\sigma) - F(\mu-\sigma) = 0.6826\ldots$.

****

It is a rather complicated matter to verify that the
integral of f over the whole region is unity, as it
should. First of all one has

$$\int_{-\infty}^{\infty} f(x)\,dx \;=\; \int_{-\infty}^{\infty} \phi(t)\,dt \;=\; \frac{1}{\sqrt{2\pi}}\int_{-\infty}^{\infty} e^{-\frac{t^2}{2}}\,dt$$

$$= \sqrt{\frac{2}{\pi}} \cdot \int_{0}^{\infty} e^{-\frac{t^2}{2}}\,dt \;\equiv\; I.$$

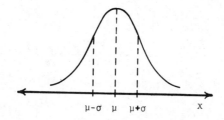

$$\mu-\sigma \quad \mu \quad \mu+\sigma \qquad x$$

Fig 9.1

Gaussian density.

Denote the last expression by I, and consider the product I^2 which can be represented by the double integral:

$$I^2 = \frac{2}{\pi} \int_0^\infty e^{-\frac{y^2}{2}} dy \cdot \int_0^\infty e^{-\frac{z^2}{2}} dz = \frac{2}{\pi} \int_0^\infty \int_0^\infty e^{-\frac{z^2+y^2}{2}} dy \, dz.$$

To evaluate this integral, change variables $z = r \sin \alpha$,
$y = r \cos \alpha$, $0 \le \alpha \le \frac{\pi}{2}$, $0 \le r < \infty$ so $dy \, dz = r \, dr \, d\alpha$,
$y^2 + z^2 = r^2$, and:

$$I^2 = \frac{2}{\pi} \int_0^\infty \int_0^{\frac{\pi}{2}} e^{-\frac{r^2}{2}} r \, dr \, d\alpha = \int_0^\infty e^{-\frac{r^2}{2}} r \, dr$$

$$= \int_0^\infty e^{-s} \, ds = 1.$$

Since $I^2 = 1$, it follows that $I = 1$, because I is
positive. This concludes the demonstration that f is a
proper density.

****

It is much easier to identify the parameters μ and
σ^2. Indeed, the following is true:

$$E(X) = \mu, \qquad var(X) = \sigma^2.$$

This follows from:

$$E(X) = \int_{-\infty}^\infty x f(x) \, dx = \int_{-\infty}^\infty (x-\mu) f(x) \, dx + \mu \int_{-\infty}^\infty f(x) \, dx$$

$$= \mu$$

by the symmetry of the normal density of f around μ.

Next, by integration by parts:

$$\text{var}(X) \; = \; \int_{-\infty}^{\infty} (x-\mu)^2 f(x) \; dx \; = \; \sigma^2 \int_{-\infty}^{\infty} y^2 \phi(y) \; dy$$

$$= \; \sigma^2 \frac{1}{\sqrt{2\pi}} \int_{-\infty}^{\infty} y^2 e^{-\frac{y^2}{2}} \; dy \; = \; \frac{\sigma^2}{\sqrt{2\pi}} \int_{-\infty}^{\infty} y \; d(-e^{-\frac{y^2}{2}})$$

$$= \; \frac{\sigma^2}{\sqrt{2\pi}} \; [-ye^{-\frac{y^2}{2}} \Big|_{-\infty}^{\infty} + \int_{-\infty}^{\infty} e^{-\frac{y^2}{2}} \; dy \;] \; = \; \sigma^2 \int_{-\infty}^{\infty} \phi(y) \; dy$$

$$= \; \sigma^2 .$$

Thus, the gaussian distribution is completely characterized by its mean and variance.

$$*********************$$

A few more interesting integrals associated with the normal distribution. For simplicity suppose that $\mu = 0$.
Then, by symmetry all moments of odd orders vanish, whereas moments of even orders are found to be:

$$\mu_{2k} \; = \; 1 \cdot 3 \cdot 5 \cdots (2k-1) \sigma^{2k}, \qquad \text{for} \quad k = 1, 2, 3, \ldots \; .$$

On the other hand, so called absolute moments of order r
(where $r > 0$, not necessarily an integer) are (again for
$\mu = 0$):

$$E(|X|^r) \; = \; \frac{2^{\frac{r}{2}}}{\sqrt{\pi}} \; \Gamma(\frac{r+1}{2}) \sigma^r, \qquad \text{for any} \quad r > 0.$$

Obviously, both expressions agree for $r = 2k$.

$$******************$$

Let X be a gaussian r.v. with mean μ and variance σ^2. Then, the pr. of the event $(a \leq X \leq b)$ can be computed from the following expression:

$$pr(a \leq X \leq b) \;=\; F(b) - F(a) \;=\; \Phi((b-\mu)/\sigma) - \Phi((a-\mu)/\sigma)$$

where F is the d.f. of X. This formula follows from the expression for F stated earlier.

Equivalently, this result can also be obtained by the analogous argument:

$$pr(a \leq X \leq b) \;=\; pr\left(\frac{a-\mu}{\sigma} \leq \frac{X-\mu}{\sigma} \leq \frac{b-\mu}{\sigma}\right) \;=\; pr\left(\frac{a-\mu}{\sigma} \leq Z \leq \frac{b-\mu}{\sigma}\right)$$

$$=\; \Phi\left(\frac{b-\mu}{\sigma}\right) - \Phi\left(\frac{a-\mu}{\sigma}\right)$$

where $Z = (X-\mu)/\sigma$.

It can be shown that the r.v. Z has standard normal distribution, with mean 0 and variance 1.

Example 1: Let $\mu = 2$, $\sigma^2 = 9$ $(\sigma = 3)$, $a = 2$, and $b = 5$. Then

$$pr(2 \leq X \leq 5) \;=\; \Phi\left(\frac{5-2}{3}\right) - \Phi\left(\frac{2-2}{3}\right) \;=\; \Phi(1) - \Phi(0)$$

$$=\; \Phi(1) - \frac{1}{2} \;=\; 0.841 - 0.500 \;=\; 0.341$$

from table.

Remember: $\Phi(z)$ is the area under the standard normal curve $\phi(z)$ from $-\infty$ up to z.

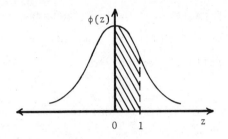

Fig 9.2

Example 1.

****

<u>Remark</u>: In applications, the gaussian approximation to pr.

of events (determined by other r.v.'s) turns out to be very

convenient for numerical computations. Its surprising

success in applications to problems in the physical sciences

justified the famous remark by Poincaré that there must be

something mysterious about the normal distribution because

"mathematicians think it is a law of nature whereas

physicists are convinced that it is a mathematical theorem."

For example, let X be a r.v. with mean $E(X) = \mu$ and

$var(X) = \sigma^2$. It is easy to check that the r.v.

$Z = (X-\mu)/\sigma$ has mean 0 and variance 1. The theorem

asserts that the distribution of Z can be approximated by

the standard gaussian distribution. Thus:

$$pr(a < X < b) \approx \Phi[(b-\mu)/\sigma] - \Phi[(a-\mu)/\sigma].$$

Exact conditions when such approximation is valid are

expressed by the famous Central Limit Theorem, which un-

fortunately we cannot discuss here. It suffices to mention

that as a first approximation, gaussian approximation is sufficiently accurate.

Example 2: i) A r.v. X is normally distributed with mean zero and variance 1. Find a number such that

$$pr(|X| > a) = \frac{1}{2} .$$

Using symmetry:

$$pr(|X| > a) = 1 - pr(|X| \le a) = 1 - 2pr(0 \le X \le a)$$

$$= 1 - 2[\Phi(a) - \Phi(0)]$$

$$= 1 - 2\Phi(a) + 2 \cdot \frac{1}{2} = 2 - 2\Phi(a) = \frac{1}{2}$$

so $\Phi(a) = \frac{3}{4}$. From the table: a = 0.675.

ii) A r.v. X is normally distributed with mean μ and variance σ^2. Find the expression for c such that

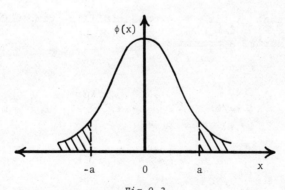

Fig 9.3

Example 2.

$$pr(|X-\mu| > c) = \frac{1}{2}$$

$$pr(|X-\mu| > c) = 1 - pr(|X-\mu| \leq c) = 1 - pr(-c \leq X-\mu \leq c)$$

$$= 1 - pr(-\frac{c}{\sigma} \leq \frac{X-\mu}{\sigma} \leq \frac{c}{\sigma})$$

$$= 1 - 2pr(0 \leq \frac{X-\mu}{\sigma} \leq \frac{c}{\sigma}) = 1 - 2\Phi(\frac{c}{\sigma}) + 1$$

$$= 2 - 2\Phi(\frac{c}{\sigma}) = \frac{1}{2}$$

so $\Phi(\frac{c}{\sigma}) = \frac{3}{4}$. Hence, $\frac{c}{\sigma} = a$ (from i), so $c = 0.675\sigma$.

<u>Example 3</u>: For a r.v. X with mean μ and variance σ^2, compute

$$pr(\mu \leq X \leq 3\mu/2)$$

assuming that X is the exponential life time. Compare with the normal approximation.

$$pr(\mu \leq X \leq 3\mu/2) = e^{-\lambda\mu} - e^{-\frac{3}{2}\lambda\mu} = e^{-1} - e^{-\frac{3}{2}}$$

$$= 0.3679 - 0.2231 = 0.1448$$

because $\lambda\mu = 1$. Now:

$$\text{pr}(\mu \le X \le 3\mu/2) \;=\; \text{pr}(0 \le \frac{X-\mu}{\sigma} \le \frac{1}{2}\frac{\mu}{\sigma}) \;=\; \text{pr}(0 \le \frac{X-\mu}{\sigma} \le \frac{1}{2}) \;,$$

because $\sigma^2 = \mu^2$ for n.e.d.

$$= \; \Phi(\frac{1}{2}) - \Phi(0) \;=\; 0.6915 - 0.5000$$

$$= \; 0.1915.$$

Section 10: When in Doubt, Approximate!

Frequently, a rough estimate of pr. of an event may be sufficient, instead of the exact value (especially, when exact calculations may be involved). The manner in which approximation is obtained depends of course on the formula to be used. However, in many situations one can obtain approximate formulae which are valid for any distributions or for a rather wide class of distributions.

10.1. One of the most useful formulae is the so called Chebyshev inequality which holds for any distribution having finite mean and (positive) variance. As an approximation, the Chebyshev inequality is rather crude, yet it gives a very handy tool both for practical and theoretical applications. We shall discuss it in two versions, both of which are of interest.

Consider first a (non-negative) life time $Y \ge 0$, with (finite) mean $m = EY$. Let g be the density of Y. Then, for any positive number $a > 0$, we can write:

$$m = \int_0^\infty yg(y)\ dy \;\geq\; \int_a^\infty yg(y)\ dy \;\geq\; a \int_a^\infty g(y)\ dy$$

$$= \; a\ pr(Y > a)$$

where the first inequality is obtained by reducing the range of integration, and the second follows from the fact that $y \geq a$ in this range; recall that $pr(Y = a) = 0$.

Hence, for any life time Y and for any $a > 0$:

$$pr(Y > a) \;\leq\; m/a.$$

This is the first version of our inequality. Clearly, it gives no information when the bound m/a is larger than 1.

Now let X be an arbitrary r.v. which may assume values of both signs, possessing a (finite) mean μ and (nonzero) variance σ^2. Clearly, a r.v. $Y = (X-\mu)^2$ is nonnegative and

$$EY \;=\; E(X-\mu)^2 \;=\; \sigma^2.$$

Hence, substituting into the above inequality for Y, one has:

$$pr((X-\mu)^2 > a) \;\leq\; \sigma^2/a.$$

Writing $a = \varepsilon^2$, and noting that $(X-\mu)^2 > \varepsilon^2$ is equivalent to $|X-\mu| > \varepsilon$ we get the famous Chebyshev inequality.

Thus, for any r.v. X with finite mean μ and positive variance σ^2, one has for an arbitrary positive ε:

$$pr(|X-\mu| > \varepsilon) \leq \sigma^2/\varepsilon^2 \qquad (\varepsilon > 0).$$

Again, the approximation is useless when σ/ε is larger than 1. It is worthwhile to note that the Chebyshev inequality gives the bound in terms of variance, of the pr. of deviation from the mean being larger than ε > 0. The pr. in question is of course given by the integral

$$pr(|X-\mu| > \varepsilon) = \int_{-\infty}^{\mu-\varepsilon} f(x)\ dx + \int_{\mu+\varepsilon}^{\infty} f(x)\ dx$$

where f is the density of X. This integral gives the exact value, but the advantage of the Chebyshev inequality is that there is no need of evaluation of these integrals.

Example: i) Suppose that X has exponential life, with mean 1/λ. Then, the exact value of the pr. of deviation, and its bound, are easily found to be:

$$pr(|X - 1/\lambda| > \varepsilon) = 1 - 2e^{-1} \sinh(\varepsilon\lambda) \leq (1/\lambda\varepsilon)^2$$

(for ε > 1/λ).

ii) For Gaussian r.v. X, it is easy to find that

$$pr(|X-\mu| > \varepsilon) = 1 - \Phi(\frac{\varepsilon}{\sigma}) + \Phi(\frac{-\varepsilon}{\sigma}) = 2[1 - \Phi(\frac{\varepsilon}{\sigma})] \leq \frac{\sigma^2}{\varepsilon^2}$$

for ε > 0.

****

<u>Note</u>: Letting $\varepsilon = k\sigma$, and taking the complement, the Chebyshev inequality may be written as:

$$pr(|X-\mu| \le \sigma k) \ge 1 - 1/k^2 \qquad \text{for } k > 0.$$

This means that the pr. is always larger than $1 - k^{-2}$ (i.e., close to 1) that X assumes values inside the interval from $\mu - k\sigma$ to $\mu + k\sigma$. This confirms that the variance is indeed a measure of the spread.

10.2. More important applications of the Chebychev ine-quality are illustrated by the following examples.

Suppose that we have several r.v.'s, say $X_1, \ldots, X_n$, and we are interested in the sum

$$S = X_1 + \ldots + X_n.$$

If individual X's are life times, S is the total life time of n items (cf. Section 5; important applications are discussed in later sections). Suppose for simplicity that the X's are independent and identically distributed (abbreviation i.i.d. is used for such a case). Then as indicated in Section 5:

$$ES = n\mu, \qquad \text{var } S = n\sigma^2$$

where μ and σ^2 are mean and variance, respectively, common to all X's.

The Chebyshev inequality gives the estimate for distribution of S:

$$pr(|S-n\mu| > \varepsilon) \leq n\sigma^2/\varepsilon^2.$$

Much more interesting is the consideration of the r.v. $\overline{X} = S/n$, that is

$$\overline{X} = (X_1 + \ldots + X_n)/n.$$

This may be regarded as some kind of average life time; in Statistics $\overline{X}$ is known as a sample mean. Hence

$$E\overline{X} = \mu, \qquad var \ \overline{X} = \sigma^2/n.$$

The Chebyshev estimate is:

$$pr(|\overline{X}-\mu| > \varepsilon) \leq \sigma^2/(n\varepsilon^2).$$

It is of great importance to observe that when $n \to \infty$, then the bound goes to zero, so

$$\lim_{n\to\infty} pr(|\overline{X}-\mu| > \varepsilon) = 0, \qquad \varepsilon > 0.$$

We express this fact by saying that $\overline{X}$ converges to μ in probability (as $n \to \infty$). This is the famous (weak) law of large numbers ("law of averages" in old fashioned terminology) for i.i.d. r.v.'s with finite variance.

****

See also a comment on normal approximation in the previous
Section

Chapter 1: Problems

1. Suppose that a life time X has density f given by:

$$
f(x) \;=\; \begin{cases} \dfrac{2}{a^2}\,(a-x) & \text{for } \; 0 \le x \le a \\[2mm] 0 & \text{for } \; x \ge a \end{cases}
$$

where a is a positive constant.

(a) Show that the d.f. F is given by

$$
F(x) \;=\; \begin{cases} (\dfrac{x}{a})\,(2 - \dfrac{x}{a}) & \text{for } \; 0 \le x \le a \\[2mm] 1 & \text{for } \; x \ge a \end{cases}
$$

(b) Show that the mean life time μ is

$$
\mu \;=\; \frac{a}{3}
$$

and that its variance is

$$
\sigma^2 \;=\; \frac{a^2}{18} \; .
$$

(c) Sketch graph of $f(x)$, $F(x)$, indicating slopes,
     etc.

(d) Take a = 2 hours, and compute the pr. that the
     life time is between 30 and 60 minutes.

2. Suppose that reading habits of a person are described
    by the following density of a time X needed to read
    a book (with the maximum time allotted to be L hours):

    $$f(t) \; = \; c \, \frac{t}{L} \, , \qquad \text{for} \quad 0 \le t \le L$$

    where c is a constant to be determined. Show that:

    (a) $c = \dfrac{2}{L}$.

    (b) The pr. of time needed being t or less is

    $$F(t) \; = \; \left(\frac{t}{L} \right)^{2} \, , \qquad \text{for} \quad 0 \le t \le L.$$

    (c) The pr. that the time needed is between a and b
         is $\dfrac{b^{2} - a^{2}}{L^{2}}$.

    (d) Moments of order n are $\mu_{n} = \dfrac{2}{n+2} \, L^{n}$
         $(n = 0,1,2,\ldots)$.

    (e) Mean reading time is $\frac{2}{3}$ L, and the variance is
         $\sigma^{2} = \dfrac{L^{2}}{18}$.

    (f) The pr. that the reading will continue an addition-
         al h hours, given that it already lasted more
         than t hours is:

    $$\frac{L^{2} - (t+h)^{2}}{L^{2} - t^{2}} \, , \qquad \text{for} \quad 0 \le h \le L - t.$$

    (g) Find the probability that the time needed to read
         the book will be less than average but more than
         half of the allotted time.

(h) Find the probability that the time needed to read
     the book will be more than average, when it is
     known that half of the allotted time has already
     elapsed.

(i) Suppose that reading of the book has been inter-
     rupted at some random instant. Show that the
     average time which would be needed to complete
     the reading (from that random instant) is (3/8)L.

3. Suppose that the life time X of a TV tube is described
    by the following density f, with the maximum life time
    equal to L years:

$$f(x) \; = \; \begin{cases} cx(L-x), & \text{for } 0 < x < L \\ 0 & \text{otherwise} \end{cases}$$

    where c is a constant to be determined. Show that

    (a) $c = 6/L^3$.

    (b) The d.f. F is given by

$$F(x) \; = \; \begin{cases} [3-2(x/L)](x/L)^2, & \text{for } 0 \le x \le L \\ 1, & \text{for } L \le x < \infty \end{cases}$$

    and F has the inflection point at $x = \frac{1}{2}L$,
    with $F(\frac{1}{2}L) = \frac{1}{2}$.

    (c) The pr. that the tube will last more than a,
         but less than b, is given by:

$$pr(a < X < b) = \frac{b-a}{L^3} [a(3L-2a-b) + b(3L-2b-a)]$$

for $0 \le a \le b \le L$.

(d) Moments of order n are:

$$\mu_n = 6L^n/(n+2)(n+3), \quad \text{for} \quad n = 0,1,2,\dots .$$

(e) Mean life time is $\frac{1}{2}L$, $var(X) = L^2/20$.

(f) The pr. that the tube will last an additional h years, given it already lasted more than x years, is:

$$\frac{F^c(x+h)}{F^c(x)} = \frac{1 - 3(\frac{x+h}{L})^2 + 2(\frac{x+h}{L})^3}{1 - 3(\frac{x}{L})^2 + 2(\frac{x}{L})^3}$$

for $h \le L-x$.

(g) Suppose that the mean life time of a tube is 5 years.

   (i) Plot graphs of f and F.
   (ii) Compute the standard deviation of the life time.
   (iii) Compute the pr. in (c) for $a = 2$ years and $b = 3$ years.
   (iv) Compute the pr. in (f) for $x = 2$ years and $h = 1$ year.

4. The scheduled length of a meeting is b (minutes, say). It is unlikely that the meeting will end before a

(minutes), and it is impossible that it will last more
than L (another meeting is scheduled in the same room).
Here $0 < a < b < L$.

Assume that the density f of the life time X
(duration of the meeting), corresponding to the above
restrictions, has the form:

$$f(x) = \begin{cases} 0 & \text{for } 0 < x < a \\ 2\beta\dfrac{x-a}{(b-a)^2} & \text{for } a < x < b \\ \dfrac{1-\beta}{L-b} & \text{for } b < x < L \\ 0 & \text{for } L < x < \infty \end{cases}$$

Here β is a constant such that $0 < \beta < 1$.

(a)   Show that the d.f. F is given by:

$$F(x) = \begin{cases} 0 & \text{for } 0 \le x \le a \\ \beta\left(\dfrac{x-a}{b-a}\right)^2 & \text{for } a \le x \le b \\ \dfrac{1-\beta}{L-b} x + \dfrac{\beta L-b}{L-b} & \text{for } b \le x \le L \\ 1 & \text{for } L \le x < \infty \end{cases}$$

(b)   Show that the average length of the meeting E(X)
is given by:

$$\mu = \frac{\beta}{3}(2b+a) + \frac{1-\beta}{2}(L+b).$$

(c) For given a, b and L, find β for which
     μ = b. Verify that indeed 0 < β < 1, and that
     always 3L - b - 2a > 0. What when β = 1 ?

(d) Take: a = 45 min., b = 50 min., L = 60 min.,
     and β = 9/10. Plot graphs of f(x) and F(x);
     indicate slope, discontinuities (if any), etc.
     Compute E(X).

(e) Find β for which E(X) = b, as in (c) above,
     using a, b and L for (d) above.

(f) Using the values from (d) above compute:
     The pr. that the meeting will terminate before
     50 minutes,
     The pr. that the meeting will terminate after
     50 minutes,
     The pr. that the meeting will terminate within
     the interval 48-51 minutes.

5. Suppose that the hazard rate of a life time X is
    given by:

$$r(t) = \frac{a}{b+t}, \qquad 0 \le t < \infty$$

where a > 1 and b > 0 are constants.
Show that:

(a) The corresponding d.f. of X is given by:

$$F(x) = 1 - \left(\frac{b}{b+x}\right)^a, \qquad 0 \le x < \infty.$$

(b)   The mean life time and variance are:

$$\mu = \frac{b}{a-1}, \qquad \sigma^2 = \frac{a}{a-2}\,\mu^2 \qquad (\text{for } a > 2)$$

$$\text{so} \quad \sigma^2/\mu^2 = a(a-2)^{-1} > 1.$$

(c)   The c.d.f. of prolongation by h is:

$$K_t(h) = 1 - \left(\frac{b+t}{b+t+h}\right)^a, \qquad 0 \le h < \infty$$

with density

$$k_t(h) = \frac{a}{b+t} \cdot \left(\frac{b+t}{b+t+h}\right)^{a+1}.$$

(d)   Take $\mu = 20$ minutes, and $a = 4$. Plot graphs of $r(t)$ and $F(x)$.

6.   Suppose that in some economics problem the hazard rate r is zero as long as life (say, income) stays below a threshold value b. At b, the rate jumps to a fixed value, and then decreases inversely proportionally to the life.

Thus, the hazard rate has the form:

$$r(x) = \begin{cases} 0 & \text{for } 0 \le x < b \\ \beta/x & \text{for } b \le x < \infty, \end{cases}$$

where β is a constant, $\beta > 1$

(a)   Compute the hazard function R for all $x \ge 0$.

(b)   Show that the corresponding life time d.f. F is given by:

$$F(x) = \begin{cases} 0 & \text{for } 0 \le x \le b \\ 1 - (b/x)^\beta & \text{for } b \le x < \infty \end{cases}$$

(Note: this is known in Economics as Pareto distribution of income).

(c) Show that the average life time is

$$\mu = \frac{b\beta}{(\beta-1)}, \quad \text{and} \quad \sigma^2 = \frac{\mu^2}{\beta(\beta-2)} \quad (\text{for } \beta > 2).$$

7. (a) Assume that the life time of a battery in a flashlight has the uniform distribution with constant density 1/L. Suppose that the mean was found to be 25 days. Determine L.

(b) Suppose that the battery has already lasted 30 days. Find the probability that the flashlight will continue to operate (with the same battery) for more than one week.

8. Suppose that a clerk made a record of time spent waiting for a bus during a week when going to work in the morning, and found that the mean waiting time was 10 minutes. Find the average time between buses and determine the probability of waiting between 5 and 19 minutes.

Consider two cases according to the assumption that the time interval between buses is distributed uniformly or exponentially.

9. In the formula for the mean waiting time w in the bus
    problem, assume that the life time distribution has the
    property that the standard deviation is proportional
    to the mean, that is $\sigma = k\mu$, where k is some
    constant.

    (a) Express w as a function of k and of μ.

    (b) Assume μ fixed, and plot the graph of w as a
         function of k. Indicate for which values of k
         the following inequality holds, $w > \mu$.

10. Suppose that inter-arrival times between buses have
    density:

    $$f(t) = \begin{cases} 2t/L^2, & 0 \le t \le L \\ 0, & L < t \end{cases}$$

    Show that the waiting time for the bus has d.f.:

    $$G(t) = \begin{cases} \frac{3}{2}\frac{t}{L}[1 - \frac{1}{3}(\frac{t}{L})^2], & 0 \le t \le L \\ 1, & L \le t \end{cases}$$

    and the mean waiting time is $w = \frac{3}{8}L$.

11. Suppose that the amount of time one spends in a bank
    is distributed with mean 10 minutes.

    (a) What is the pr. that a customer <u>arriving at
         random</u>, will spend more than 10 minutes in
         the bank?

(b) What is the average time in the bank, for the
     customer in (a) ?

Consider three cases separately, according to
distribution of the amount of time being

(1) uniform

(2) exponential

(3) same as in Problem 5.

Give numerical results.

12. Suppose that the amount of time that a light bulb
     works before burning itself out is uniformly distrib-
     uted with mean μ. Suppose that a person enters a
     room in which a light-bulb is burning for some time.
     What is the pr. that he/she will be able to use that
     light for more than the average μ ?

13. Suppose that independent life times X and Y have
     exponential distributions, but with <u>different</u> parameters,
     λ and μ respectively.

     (a) Show that the life time $Z = X + Y$ has density
          g of the form:

          $$g(z) = \frac{\lambda\mu}{\mu-\lambda} (e^{-\lambda z} - e^{-\mu z}) \qquad \text{for } \lambda \neq \mu.$$

     (b) Verify that the d.f. G of Z is

          $$G(z) = 1 - \frac{\mu}{\mu-\lambda} e^{-\lambda z} + \frac{\lambda}{\mu-\lambda} e^{-\mu z} \qquad \text{for } \lambda \neq \mu.$$

(c) Verify that $EZ = EX + EY = 1/\lambda + 1/\mu$.

(d) Deduce that for $\lambda = \mu$, the above expressions
     reduce to that in Example 1, Section 5.

14. Suppose that the joint density of two life times X
     and Y has the form:

$$f(x,y) \;=\; \begin{cases} c(x+y) & \text{for } x, y \text{ in region } T \\ 0 & \text{for } x, y \text{ outside } T \end{cases}$$

     where T is a triangle in the first quadrant deter-
     mined by lines:

$$x = a, \qquad y = 0, \qquad y = mx$$

     where a and m are positive constants.

(a) Show that $c = 6a^{-3}(2m+m^2)^{-1}$.

(b) Evaluate $\operatorname{pr}(\frac{1}{2}a \le X \le a,\; 0 \le Y \le \frac{1}{2}am)$.

(c) Find marginal densities $f_1(x)$ and $f_2(y)$, and
     verify that X and Y are dependent.

15. In this problem the life time X has density f of
     the form given in Problem 3. Suppose that the gain
     from operating a certain equipment depends on the
     duration of its life time in the following way. In
     the initial stages of operation when life does not
     exceed a fixed length a, gain increases proportion-

ally to the length of life. Then when operation is established, gain remains constant throughout the remaining period from a to L.

More precisely, the cost function has the following form:

$$\varphi(x) = \begin{cases} mx & \text{for} \quad 0 \le x \le a \\ ma & \text{for} \quad a \le x \le L \end{cases} \quad \text{where} \quad \begin{array}{l} 0 \le a \le L \\ 0 \le m < \infty \end{array} \quad \begin{array}{l} \text{are} \\ \text{constant} \end{array}$$

(a) Show that the average gain is:

$$E\varphi(X) = \frac{1}{2} ma[2 - 2(a/L)^2 + (a/L)^3].$$

(b) Suppose now that the slope m is related to the position a in a manner described by the following relation:

$$ma = c(5a+4L) \qquad \text{for} \quad 0 \le a \le L,$$

where c is a fixed constant (measured in dollars per unit of life).

Regard the average gain $E\varphi(X)$ as a function of the location a, and show that gain is maximum for the following choice of a: $a_m = \frac{1}{2}L$, and the value of this maximum is $(169/32)cL$.

Hint: Use the following factorization of the cubic:

$$20z^3 - 18z^2 - 16z + 10 = 2(5z^2-2z-5)(2z-1)$$

for $0 \le z \le 1$.

(c)   Select L = 10 years, and plot the following graphs
       as functions of a: (take c = 2 hundred dollars
       per year) Hint: for convenience use: z = a/L.

       (i)   ma; for selected values of a, indicate
              the corresponding cost function.

       (ii)   m; what are values of m at a = 0,
              $a = \frac{1}{2} L$, a = L ?

       (iii) Average gain Eφ(X), indicating the maximum.

16. Suppose that the life time X has density f of the
     form given in Problem 3. Suppose that when the life
     time is smaller than some value t, where $0 < t \leq L$,
     then the cost of operation is constant but inversely
     proportional to t.

          If the life time is larger than t, the cost is
     zero. Thus, the cost function is

$$\varphi(x) \;=\; \begin{cases} \dfrac{cL}{t} & \text{for } 0 \leq x \leq t \\[2ex] 0 & \text{for } t < x \leq L \end{cases}$$

where c is a positive scaling constant.

(a)   Show that the average cost is:

$$\mathbb{E}\,\varphi(X) \;=\; c\left(\frac{t}{L}\right)\left(3 - 2\,\frac{t}{L}\right) .$$

(b)   Let now t be a variable. Plot the graph of the
       average cost $\mathbb{E}\varphi(X)$ as a function of t, and
       show that it has a maximum at t = (3/4)L, equal
       to (9/8)c.

17. Maintaining cost of a certain equipment depends on the duration of its life in the following way. In the initial stages of operation when life does not exceed a fixed length a, cost increases sharply, proportionally to the square of life. Then when operation is established cost remains constant throughout the whole period from a to some other point b. When equipment is old, that is when it life is over b, cost increases rather slowly with constant rate.

More precisely, the cost function has the following form:

$$\varphi(x) = \begin{cases} \lambda^2 x^2 & \text{for } 0 \leq x \leq a \\ \lambda^2 a^2 & \text{for } a \leq x \leq b \\ m(x-b) + \lambda^2 a^2 & \text{for } b \leq x < \infty \end{cases}$$

$$a < b, \quad m > 0, \quad \lambda > 0.$$

Furthermore, it is assumed that life time X is exponential, with mean $\mu = 1/\lambda$.

(a) Draw the graph of the cost function.

(b) Show that the average cost is

$$E\varphi(X) = 2 - 2(\lambda a + 1)e^{-\lambda a} + \frac{m}{\lambda}e^{-\lambda b}.$$

(c) Suppose now that the period of the constant cost operation has fixed duration h, so b = a + h.

Regard the average cost $E\varphi(X)$ as a function of the location a, and show that the cost is minimum for the following choice of a:

$$a = \frac{m}{2\lambda^2} e^{-\lambda h}$$

and the value of this minimum cost is $2(1-e^{-\lambda a})$, independent of length h.

(d) Select the average life $\mu = 10$ days, $h = 20$ days, and take $m = \frac{1}{2}$. Plot the graph of the cost $E\varphi(X)$ as the function of a, indicating the position of the minimum.

18. Suppose that the life time X has the density f given in Problem 1, with mean $\mu = \frac{a}{3}$.

Suppose that the cost of operation of the system is such that whenever the life stays below its mean value, the cost is proportional to the length of life, plus a fixed initial cost.

Whenever the life overshoots its mean value, the cost remains constant and equal to its value when life is equal to its mean.

In symbols, the corresponding cost function φ is given by:

$$\varphi(x) = cx + b \quad \text{for} \quad 0 \leq x \leq \mu,$$
$$\varphi(x) = c\mu + b \quad \text{for} \quad \mu \leq x \leq a,$$

where $c \geq 0$ is the cost per unit of time and
$b \geq 0$ is a fixed initial cost.

(a) Show that the average cost $E\varphi(X)$ is given by:

$$E\varphi(X) = \frac{19}{27} c\mu + b.$$

(b) Take $\mu = 9$ hours, $c = 6$ dollars per hour,
$b = 14$ dollars. Compute $E\varphi(X)$.

(c) (i) For fixed c and b , plot the graph of the
average cost $E\varphi(X)$ as a function of the
mean μ .

(ii) Plot the graph of the cost function $\varphi(x)$,
as a function of x .

19. An old lady noticed that when her favorite tea pot is
younger than the average life of tea pots, it produces
good tea. If, however, its life is larger than the
average, then the taste of tea deteriorates proportionally to the age of the tea pot. Express the lady's
discomfort by the "cost function" $\varphi(x)$ which equals
0 when $x < \mu$, and equals ax when $x \geq \mu$, where
a is constant, and μ is the mean.

Assuming that tea pot life is exponential, show
that the lady's average discomfort $E\varphi(X)$ equals
$2a\mu e^{-1}$.

20. Consider a situation when an animal contacts a dis-
    ease which incapacitates it for a fixed length $h > 0$
    of time; afterwards the animal becomes immune. Sup-
    pose that the animal's incapacity stays constant during
    the illness (measured in some convenient units), whose
    magnitude increases with the animal's age   a   at the
    time of infection.

    Thus the "illness function" is given by

    $$\varphi(x) = ca \qquad \text{for} \quad a < x < a+h$$

    and $\varphi(x) = 0$ otherwise. Here c is a positive
    constant.

    Assume (for simplicity) that the animal's life
    is exponential.

    Compute the "average illness," and show that it
    reaches a maximum when the infection instant equals
    the average life time. Sketch a graph of the "average
    illness" as a function of a. What would be the
    meaning of $h \to \infty$?

21. Suppose that the density f of a life time X is
    <u>symmetric</u> around a point $c > 0$, that is: $f(c-t) = f(c+t)$ for all t such that $0 < t < h$ where h
    is a constant such that $c-h > 0$, and $f = 0$ outside
    the interval $(c-h, c+h)$.

    Show that the mean EX equals c.

22. Suppose that the cost function φ has the form:

$$\varphi(x) = \begin{cases} a & \text{when } x \le t \\ b & \text{when } x > t \end{cases}.$$

Show that the average cost is $aF(t) + bF^c(t)$. Plot its graph as a function of t. What if $a = b$? Is there any maximum or minimum?

23. Let f and F be the density and d.f., respectively, of a life time X. Consider the cost function defined by: $\varphi(x) = F(x)$. Verify that $\mathbb{E}\varphi(X) = \frac{1}{2}$ always!

24. Let R be a hazard function, and let X be a life time with distribution determined by R. Consider the cost function defined by $\varphi(x) = R(x)$. Verify that $\mathbb{E}\varphi(X) = 1$ always.

25. Suppose that the hazard rate r has the form:

$$r(t) = \begin{cases} c \cdot \sin(\frac{\pi}{L}t) & \text{for } 0 \le t \le L \\ c & \text{for } L < t < \infty \end{cases}$$

where L and c are positive constants.

Show that the d.f. F of the life time has the form

$$F(x) = 1 - e^{-R(x)}$$

where

$$R(x) = \begin{cases} \dfrac{cL}{\pi} \left(1 - \cos \dfrac{\pi x}{L}\right) & \text{for } 0 \le x \le L \\[2ex] \dfrac{2cL}{\pi} + c(x-L) & \text{for } L \le x < \infty. \end{cases}$$

Let the cost function $\varphi = 0$ for $t < L$, and $\varphi = a$ for $t > L$ where $a = \dfrac{2cL}{\pi}$. Show that the average cost is ae^{-a}.

26. A manufacturer produces two models of an apparatus, say Model A and Model B. Let X be the life time of Model A, and Y the life time of Model B. For simplicity, assume that X and Y are independent exponential lifes with parameters λ and μ, respectively.

(a) In order to compare these models, the manufacturer is interested in the ratio Y/X of their life times. Show that for $z \ge 0$:

$$\mathrm{pr}(Y/X \le z) = \mu z / (\mu z + \lambda).$$

What is the average ratio $E(Y/X)$? Is there anything strange about it? Explain.

(b) A comparison of models may be based on the pr. that Model A lasts longer than Model B. Show that

$$\mathrm{pr}(Y \le X) = \mu / (\mu + \lambda).$$

(c) Suppose that from the records of sales it was estimated that the average life of Model B is 3 times larger than the average life of Model A. Compute the above probability in this case.

(d) Comment on agreement or discripancy between results obtained in (c).

27. Let X be the life time of a new device, and let Y be the life time of an old device. Assume X and Y are independent and identically distributed with the common d.f. F with density f. In order to compare these devices, it is required to evaluate the probability that Y > X, but with the restriction that Y ≤ a, where a is fixed.

(a) Write down the joint density f(x,y) of X and Y.

(b) Indicate the region in the plane corresponding to the event X < Y ≤ a.

(c) Set up the double integral for the required probability.

(d) Evaluating this integral, show that its value is $(1/2)[F(a)]^2$, for arbitrary F.

28. Let X and Y be two independent life times, each having exponential distribution with parameter λ and

μ, respectively. Assume $\lambda \neq \mu$ (for $\lambda = \mu$ see Section 6).

(a) Show that the life time $Z = \min(X,Y)$ has also exponential distribution but with parameter $(\lambda + \mu)$.

Deduce that $E(Z) \leq \frac{1}{2}[E(X) + E(Y)]$

(b) Show that the life time $T = \max(X,Y)$ has d.f. G of the form:

$$G(t) = 1 - e^{-\lambda t} - e^{-\mu t} + e^{-(\lambda+\mu)t}, \qquad t \geq 0$$

with density

$$g(t) = \lambda e^{-\lambda t} + \mu e^{-\mu t} - (\lambda+\mu)e^{-(\lambda+\mu)t}, \quad t \geq 0$$

and moments

$$ET^n = n! \left[\frac{1}{\lambda^n} + \frac{1}{\mu^n} - \frac{1}{(\lambda+\mu)^n} \right], \qquad n = 0,1,\dots .$$

29. A car owner installed two new tires of different brands. Let X be the life time of the first tire assumed to have uniform distribution over $(0,a)$. Let Y be the life time of the second tire assumed to have uniform distribution over $(0,b)$. Assume that X and Y are independent. Suppose also that $b < a$.

(a) Write down the joint density $f(x,y)$ of X and Y.

(b) Plot the rectangle in the plane in which X and
     Y take their values jointly, and indicate the
     region corresponding to the event (Y > X).

(c) Show that the probability of Y being larger than
     X is b/(2a).

(d) Compute this probability when it is known that the
     mean of X is twice the mean of Y.

30. Let X be the life time of an engine in a car, and let
     Y be the life time of the body of a car. Assume (for
     simplicity) that X and Y are independent life times
     having n.e.d. with parameters λ and μ, respectively.

          As long as the engine is OK (i.e., when X ≥ Y),
     one can assume that eventual damage to the body of the
     car is expressed by a fixed number c (in dollars, say).
     On the other hand, when X < Y the utility of the car
     drops down to zero. In other words, the utility func-
     tion φ is of the form:

$$\varphi(x,y) \;=\; \begin{cases} c & \text{when } x \ge y \\ 0 & \text{when } x < y \end{cases} \qquad (x \ge 0, \ y \ge 0, \ c > 0)$$

Show that the average utility is:

$$E\varphi(X,Y) \;=\; c \iint_{x \ge y} \lambda e^{-\lambda x}\mu e^{-\mu y}\, dx\, dy \;=\; c\mu/(\lambda+\mu).$$

31. A person must go to a drug store and to a bookshop;
     let X and Y be time spent in these two places,
     respectively.   Assume that X and Y are indepen-
     dent r.v.'s having the same n.e.d. with a parameter
     λ.

        Suppose that this person can do this shopping
     only if the total time spent, X + Y, is t or less
     (where t is fixed); imagine then that the person
     receives a reward c under such conditions. When
     X + Y is larger than t, the reward is zero (due to
     missing the next appointment, say). That is, the
     reward function φ is of the form:

$$\varphi(x,y) \;=\; \begin{cases} c & \text{when } x+y \le t \qquad (c>0) \\ 0 & \text{when } x+y > t \end{cases}$$

Show that the average reward is:

$$E\,\varphi(X,Y) \;=\; \lambda^2 c \iint_{x+y \le t} e^{-\lambda(x+y)}\,dx\,dy$$

$$=\; c[1 - (1+\lambda t)e^{-\lambda t}]$$

and verify that this is an increasing function of t.

32. The error in a certain angle-measuring device has been
     found to be normally distributed with mean one minute
     and standard deviation σ of three minutes.

     (a) What is the pr. that a measurement is in error by
          more than three minutes?

(b) Repeat if one minute is subtracted from the measured value.

33. The lifetimes of two competing brands of vacuum tubes
can be viewed as normally distributed r.v.'s. Brand A
has mean life of 27 hours and a standard deviation of
5 hours, whereas brand B has $\mu = 30$ and $\sigma = 2$ hours.

(a) Which brand should be chosen for use in an experi-
mental aircraft with a flight time of 30 hours?

(b) Same question for 34 hours.

(c) Show that the pr. of a negative lifetime for each
brand is negligible (this can be interpreted as
the pr. that a tube is no good on delivery).

34. Assume that the life in hours of a flashlight is normally distributed with a mean of 100 hours. If a purchaser requires at least 90% of them to have lives
exceeding 80 hours, what is the largest value that
the standard deviation σ can have and still have the
purchaser satisfied?

35. (a) Show that for the standard normal distribution:

$$\Phi(-x) = 1 - \Phi(x), \quad \text{for any } x \geq 0.$$

Show this relation on the graph of density ϕ .

(b) Let X be a Gaussian r.v. with mean μ and
standard deviation σ. Show that

$$pr(|X| > a) = 2 - \Phi((a+\mu)/\sigma) - \Phi((a-\mu)/\sigma)$$

for any positive a. Indicate this relation on
the graph of density of X.

36. Suppose that measurements in a certain experiment can
be represented by a Gaussian random variable of the
form $X = c + E$, where c is a constant value and E
is the Gaussian error with mean 0 and variance σ^2.

(a) Show that the mean and the variance of X are c
and σ^2, respectively.

(b) Show that the pr. of good performance, i.e. the pr.
that the (absolute) error does not exceed some
$\varepsilon > 0$, is:

$$pr(|E| \leq \varepsilon) = 2\Phi(\varepsilon/\sigma) - 1.$$

(c) Verify that $\varepsilon < \sigma$ implies that this pr. is less
than 0.682.

37. In a certain equipment there is a critical value c of
the life time X of the equipment, such that when
fluctuation of X around c does not exceed (in absolute value) c itself, then equipment performs well.
(Here $c > 0$ is a constant).

Suppose that X has d.f. F, and mean μ and variance σ^2.

(a) Compute the pr. of good performance, i.e. show that

$$pr(|X-c| \leq c) = F(2c).$$

(b) Show that the Gaussian approximation to this pr. is

$$\Phi(\frac{2c-\mu}{\sigma}) + \Phi(\frac{\mu}{\sigma}) - 1.$$

(c) Assuming that $\mu = 2c$ and $\sigma^2 = c^2$, show that the pr. in (b) equals 0.477 ... .

38. Let X be a Gaussian r.v. with mean μ and variance σ^2. Consider the following life time:

$$Y = |X - \mu|$$

(absolute fluctuation around the mean). Show that Y has "folded Gaussian" density given by

$$f(y) = \sqrt{\frac{2}{\pi}} \frac{1}{\sigma} e^{-\frac{y^2}{2\sigma^2}}, \qquad \text{for } y > 0$$

and $f = 0$ for $y < 0$.

Verify that:

$$E(Y) = \sqrt{\frac{2}{\pi}} \sigma, \qquad var(Y) = \sigma^2(1 - \frac{2}{\pi}).$$

39. Let X be a Gaussian r.v. with mean μ and variance σ^2. Consider the following life time:

$$U \;=\; (X-\mu)^2$$

(square variation around the mean); Show that U has "chi-square" density given by

$$f(u) \;=\; \frac{1}{\sqrt{2\pi u}\,\sigma}\, e^{-\frac{u}{2\sigma^2}}, \qquad \text{for}\quad u > 0$$

and $f = 0$ for $u < 0$.

Verify that:

$$E(U) \;=\; \sigma^2, \qquad var(U) \;=\; 2\sigma^4, \qquad E(\sqrt{U}) \;=\; \sqrt{\frac{2}{\pi}}\;\sigma.$$

40. Let X and Y be two independent Gaussian r.v.'s with mean μ_1, μ_2 and variance σ_1^2, σ_2^2, respectively. Show that a r.v. $Z = X + Y$ is also Gaussian with mean $\mu_1 + \mu_2$ and variance $\sigma_1^2 + \sigma_2^2$.

(Hint: see Sections 5 and 9; complete the square in convolution integral.)

41. Suppose that the life time X has a d.f. F of the form:

$$F(t) \;=\; 1 - \sum_{k=1}^{n} a_k e^{-\lambda_k t}, \qquad t \geq 0$$

where $a_k > 0$, $\lambda_k > 0$ and

$$\sum_{k=1}^{n} a_k \;=\; 1.$$

(i) Verify that $F(t)$ is a d.f., with $F(0) = 0$,
$F(\infty) = 1$, with density

$$f(t) = \sum_{k=1}^{n} a_k \lambda_k e^{-\lambda_k t}, \qquad t > 0.$$

(ii) Show that the average life time is:

$$E(X) = \sum_{k=1}^{n} a_k / \lambda_k.$$

(iii) Show that the Laplace transform is:

$$f^*(\alpha) = \sum_{k=1}^{n} \lambda_k a_k / (\lambda_k + \alpha).$$

(iv) Take $n = 2$, and plot graphs of $F(t)$ and $f(t)$.

42. Let X be a life time with mean μ, variance σ^2 and a d.f. F. Let X_1 and X_2 be two independent life times with the same distribution as X. Consider

$$Z = \min(X_1, X_2).$$

Show that

$$\int_0^\infty F(x) F^c(x)\, dx = \mu - m > 0$$

where $m = E(Z)$.

43. Consider the extreme life times M_+ and M_- defined in Section 6 for i.i.d. life times, with the common life time X. Show analytically that always:

$$E(M_-) \leq E(X) \leq E(M_+).$$

44. Show that moments of the life time X with d.f. F
    are given by:

$$E(X^n) = n \int_0^\infty x^{n-1} F^c(x) \, dx, \qquad n = 1, 2, \ldots .$$

Chapter 2

Be Discreet with Discrete

We shall now discuss life times which assume values in a countable set only (that is finite or denumerably infinite). We shall call them "discrete life times." This may be for example the waiting time measured in fixed units, like 1, 2, 3 and so on, hours, days, etc. We can also make observations at specific instants of time, say at noon every day.

In this category we may incorporate other variables not usually designated as life times, like, for example, a number of objects in some collection. The number of people waiting in a line, the number of patients in a hospital, the number of engaged machines in a shop, and many other examples of this kind, are typical illustrations.

The main difference from the previously considered situation is that we shall not use integration (or differentiation), but rather summation (finite or infinite). In this respect our discussion will be simpler, although some of our sums may be rather strange.

This Chapter deals with three main discrete distributions, namely binomial, geometric and Poisson -- they corre-

sond to three well known series from calculus. On the application side, we are in the heart of Probability Theory.
Bernoulli trials (Sections 11-13) go back to James Bernoulli,
a 17th century Swiss mathematician; and the Poisson distribution originated with Simeon D. Poisson, a French mathematician working in the first half of the 19th century. It
is amazing how many modern applications fit well into these
venerable formulae.

   We shall see here many applications to diverse fields,
ranging from simple and amusing to more complicated and serious examples. We shall also hint at some modern develop-
ments which lie behind surface (see Section 15, and some
Problems).

Section 11:   Bernoulli Trials

11.1.

   We shall begin with a rather important case of Bernoulli
trials which find an extremely large number of applications.
Suppose that we perform an "experiment" which may result
only in two outcomes. This of course need not be a physical
experiment, but any observation, situation, or action in
which only two aspects are of interest. For example, taking
an exam one may pass or fail, a signal may be on or off, in
a game one can win or lose, and so on; we need not mention
the well known coin tossing! For convenience, any such
attempt or experiment will be referred to as a <u>trial</u>. Its
outcomes are conveniently called a success (S) and a failure

(F); what is S and what is F is immaterial (one man's success is another man's failure).

We shall consider an event S (that success occured) and an event F (the failure took place), and in agreement with our procedures (Section 1) associate probabilities with these events:

$$Pr(S) = p, \qquad pr(F) = q,$$

where $p + q = 1, \quad 0 \leq p \leq 1, \quad 0 \leq q \leq 1.$

The condition $p + q = 1$ is in agreement with the fact that S and F are two complementary events.

Now, we consider an experiment which consists of repeating such trials n times. This may be, for example, n exams to be taken, n machines in a shop, n sources sending signals, etc. Each of these n trials may result either in a success or in a failure. The big problem is to find the probability that n trials will result in exactly j successes, where j ranges from 0 to n. If n is a number of chairs in the room, and success means that a chair is occupied (i.e., there is a customer in the room), then our problem is to have the expression for the probability that there will be exactly j chairs occupied; if j = 0, all chairs are free, if j = n, all chairs are taken.

To be more specific, we must impose further assumptions which will specify <u>Bernoulli</u> <u>trials</u>:

i)   all trials have the same pr. p of resulting in
     success S,

ii) all trials are independent of each other.

     The assumption of independence is essential, and it
simply means that, say, the pr. of two successes SS is
simply p^2, pr. of FF is q^2, pr. of SF is pq, and
so on (this is analogous to notion of independence discussed
in Section 5).

     Let X be a "life time" (i.e., a random variable)
representing a number of successes in n trials. We shall
write for its distribution, that is the probability:

$$pr(X = j) = P(j), \qquad j = 0, 1, \ldots, n.$$

Since the events of getting 0 successes, 1 success, ...,
j successes are necessarilly exclusive the pr.'s add to
unity:

$$\sum_{j=0}^{n} P(j) = P(0) + P(1) + \ldots + P(n) = 1.$$

It is our task to find the formula for P(j). For this purpose we shall examine events (X = j) successively for each
j, as in the following table. Remember, there are n
trials and we look for all possible arrangements of successes
S.

X = j	sequence of length n	pr. of each sequence	P(j) = pr(X=j)
j = 0	FFF........F	q^n	q^n
j = 1	SF.........F F.....S....F F.........FS	pq^{n-1} each	npq^{n-1}
j	F..S..FS..F..S j S's n-j F's	$p^j q^{n-j}$ each	$\binom{n}{j} p^j q^{n-j}$
j = n	SSS........S	p^n	p^n

Indeed, if in a string of n trials there is exactly j S's, then there must be exactly n - j F's; the pr. of getting j S's is p^j, by independence; the pr. of getting n-j F's is q^{n-j}; hence, the pr. of a sequence with j successes is (again by independence) $p^j q^{n-j}$.

However, j successes are distributed among n places -- the number of selections of j places out of n, that is a number of sequences of length n carrying j S's and n-j F's, is given by the binomial coefficient

$$\binom{n}{j} = \frac{n!}{(n-j)!\,j!} \, ,$$

where ! denotes factorial:

$$k! = 1 \cdot 2 \cdots (k-1)k = k \cdot (k-1)!$$

with 0! = 1.

Since each of such sequences carrying j S's is different (sequences being exclusive), the pr.'s add, and consequently the pr. of the composite event $(X = j)$ is

$$P(j) = \binom{n}{j} p^j (1-p)^{n-j}, \qquad j = 0,1,\ldots,n.$$

This expression is known as the <u>binomial</u> <u>distribution</u>, with parameters n,p. Observe that it agrees with values of j in the table.

11.2.

We shall look closely at the above expression for $P(j)$. First of all recall a few properties of the binomials coefficients:

$$\binom{n}{j} = \binom{n}{n-j}, \qquad \binom{n}{j} = \frac{n}{j}\binom{n-1}{j-1}, \qquad \binom{n}{j} + \binom{n}{j-1} = \binom{n+1}{j}.$$

For small values of n, the binomial coefficients are easily obtained from the so called Pascal triangle, as it is well known.

The name "binomial" comes from the binomial expansion:

$$(a+b)^n = \sum_{j=0}^{n} \binom{n}{j} a^j b^{n-j}$$

for any real a and b.

Taking $a = p$ and $b = q$, it follows immediately that $\sum_{j=0}^{n} P(j) = (p+q)^n = 1$.

Letting $a = b = 1$, one finds that the sum of all binomial coefficients is: $\sum_{j=0}^{n} \binom{n}{j} = 2^n$. Letting $b = 1$ and $a = -1$, one has:

$$\sum_{j=0}^{n} \binom{n}{j}(-1)^j = 0.$$

Consider now the ratio:

$$\frac{P(j)}{P(j+1)} = \frac{j+1}{n-j}\frac{q}{p}.$$

It follows that according to values n and p, the following possibilities may occur: $P(j)$ as a function of j may increase, may decrease, or may increase first and then decrease.

The expected number of successes $E(X)$ is defined by

$$E(X) = \sum_{j=0}^{n} j \cdot P(j).$$

Similarly, the expected cost (cf. Section 7) is

$$E\varphi(X) = \sum_{j=0}^{n} \varphi(j)\, P(j).$$

It can be shown that

$$\boxed{E(x) = np \quad \text{and} \quad \text{var}(X) = npq.}$$

Indeed:

$$E(X) = \sum_{j=0}^{n} j\binom{n}{j}p^j q^{n-j} = \sum_{j=0}^{n} j\, \frac{n!}{j!\,(n-j)!}\, p^j q^{n-j}$$

$$= n \sum_{j=1}^{n} \frac{(n-1)!}{(j-1)!\,(n-j)!}\, p^j q^{n-j} = np \sum_{j=1}^{n} \binom{n-1}{j-1}p^{j-1}q^{n-j}$$

$$= np \sum_{k=0}^{n-1} \binom{n-1}{k}p^k q^{n-1-k} = np \qquad (k = j-1).$$

120   *Be Discreet with Discrete*

In order to find var X, it is more convenient to compute EX(X-1) first:

$$EX(X-1) = \sum_{j=0}^{n} j(j-1) \frac{n!}{j!(n-j)!} p^j q^{n-j}$$

$$= n(n-1) \sum_{j=2}^{n} \frac{(n-2)!}{(j-2)!(n-j)!} p^j q^{n-j} = n(n-1)p^2$$

so $EX^2 = EX(X-1) + EX = n(n-1)p^2 + np = n^2p^2 + npq.$
Hence var X $= EX^2 - (EX)^2 = n^2p^2 + npq - (np)^2 = npq.$

****

Recall that P(j) is the pr. of exactly j successes. In order to find the pr. of j or less successes, one has

$$pr(X \leq j) = \sum_{i=0}^{j} P(i) = P(0) + P(1) + \ldots + P(j).$$

This is clearly an increasing function of j. Obviously, $pr(X \leq n) = 1.$ Similarly, the complementary pr. of more than j successes is

$$pr(X > j) = \sum_{i=j+1}^{n} P(i) = P(j+1) + \ldots + P(n).$$

This is analogous to the d.f. F in the continuous case, except that now the argument is discrete:

$$\boxed{F(j) = \sum_{i=0}^{j} \binom{n}{i} p^i q^{n-i}, \qquad 0 \leq j \leq n}$$

Note: Special cases of interest:

(i) $p = q = 1/2$; $P(j) = \binom{n}{j} \dfrac{1}{2^n}$.

(ii) $p = 1$; $P(n) = 1$, $P(j) = 0$, $j \neq n$.

(iii) $p = 0$; $P(0) = 1$, $P(j) = 0$, $j \neq 0$.

11.3.

There is another approach to Bernoulli trials which is very instructive. Let Z be a r.v. which assumes values 0 and 1, only, with probabilities:

$$pr(Z = 0) = q, \qquad pr(Z = 1) = p.$$

Hence, $E(Z) = p$ and $var(Z) = pq$. Clearly, Z represents the result of a trial (1 stands for success S and 0 for failure F). Hence, r.v.'s $Z_1, Z_2, \ldots, Z_n$ which are i.i.d. (independent identically distributed) with common distribution being that of Z, represent results of consecutive n trials. The number of successes in n trials which we now denote by X_n (to stress dependence on n) is therefore:

$$X_n = Z_1 + \ldots + Z_n.$$

As noted in Section 5 and also in Section 10.2, the mean and variance of X_n are simply:

$$E(X_n) = np, \qquad var(X_n) = npq,$$

in agreement with our earlier calculations. Indeed, our evaluation of $P(j)$ is just the determination of distribution of X_n, in the manner of combination of life times as discussed in Section 5.

Let us now consider the sample mean of successes in n trials:

$$X_n/n = (Z_1 + \ldots + Z_n)/n.$$

Clearly, X_n/n has the same mean as Z (although these r.v.'s are different), but its variance is $\text{var}(X_n)/n^2$ and will decrease with n:

$$E(X_n/n) = p, \qquad \text{var}(X_n/n) = pq/n.$$

By the Chebyshev inequality from Section 10.2 we have:

$$\text{pr}(|X_n/n - p| > \varepsilon) \leq pq/(n\varepsilon^2), \qquad \varepsilon > 0.$$

As $n \to \infty$, the bound goes to 0, so the indicated pr. tends to zero. We thus obtained the Bernoulli Law of Large Numbers which asserts that (with limit in probability, as in Section 10):

$$\lim_{n \to \infty} \frac{Z_1 + \ldots + Z_n}{n} = p.$$

Thus, as n becomes large the sample mean X_n/n for Bernoulli trials converges (in probability) to the "population mean" p (which coincides here with the pr. of a success).

This result contains a mathematical counterpart of the intuitive notion of probability as frequency. The sample mean X_n/n --actually its observable values -- yield the frequency of occurrence of a success, and it tends to p when the number of observations n becomes large. Compare this statement with our earlier remarks on the frequency approach in Section 1.

Section 12: Applications of Binomial

The binomial distribution discussed in the previous section occurs frequently in applications, and this section lists some illustrative examples.

12.1. Sampling with replacement

Suppose that there are M objects (coins, balls, machines, cars) of which m are of a special kind in which we are interested; label them as "good" objects.

An object chosen "at random," inspected and its kind noted, and then it is returned. The procedure is repeated n times (where, of course n may be larger than M); in other words, a sample of size n is drawn with replacements from a population of size M. What is the pr. that this sampling will produce exactly j "good" objects?

Clearly, we have here the situation of n Bernoulli trials, "at random" meaning that each object has equal chance to be chosen, so p = m/M. Hence, the required pr. is:

$$P(j) \; = \; \binom{n}{j}(m/M)^j(1 - m/M)^{n-j}, \quad j = 0,1,\ldots,n$$

and the average number of good choices is nm/M.

As an example, what is the pr. that in a sequence of 7 digits (as in a telephone number) there will be at least two digits 5 ? Since there are 10 digits, M = 10, and 5 being one of them m = 1, so p = 1/10 obviously. Since n = 7, the required pr. is:

$$\sum_{j=2}^{7} P(j) \; = \; 1 - P(0) - P(1)$$

$$= \; 1 - (9/10)^7 - 7(9/10)^6(1/10) \; = \; 0.17.$$

In this example the mean is np = 7/10 and variance is npq = 63/100. To compute the above expression we may use Gaussian approximation as in Section 9.

12.2. Overselling of tickets

An airline knows that on the average r of the people making reservations on a certain flight will not show up. Consequently, the policy is to sell t tickets for a flight that can only carry s passengers (t > s).

What is the pr. that there will be a seat available for every passenger that shows up? Let X be the number of passengers showing up; clearly, X assumes values from 0 to t. Hence, EX = t - r, and therefore p = 1 - r/t.

The pr. of exactly j passengers showing up is the binomial

$$P(j) \;=\; \binom{t}{j}p^{j}(1-p)^{t-j}.$$

Consequently, the required pr. is:

$$pr(X \le s) \;=\; \sum_{j=0}^{s} P(j).$$

In the special case when only one seat is oversold, then this pr. is $1 - (1 - r/t)^{t} \approx 1 - e^{-r}$, as can be seen by putting s = t - 1.

As an example, take r = 10, t = 100, and s = 95. Then p = 0.9. The mean is tp = 90 and variance tp(1-p) = 9. Hence, the Gaussian approximation yields:

$$pr(0 \le X \le 95) \;=\; \Phi(\frac{95-90}{3}) - \Phi(\frac{0-90}{3}) \;=\; \Phi(\frac{5}{3}) - \Phi(-30)$$

$$\approx \Phi(\frac{5}{3}) \;=\; 0.952.$$

In another special case suppose that the average number of passengers showing up equals the number of seats; that is t - r = s. Then, p = s/t and variance is s(1 - s/t). Hence the Gaussian approximation yields:

$$pr(0 \le X \le s) \;=\; \Phi(\frac{s-s}{\sigma}) - \Phi(\frac{-s}{\sigma}) \;\approx\; \Phi(0) \;=\; \frac{1}{2}$$

which looks rather bad for a passenger.

12.3. <u>ESP</u>.

An individual claims to have extrasensory perception
(ESP). As a test, a fair coin is tossed n times and he
is asked to predict in advance the outcome.

Our individual gets k out of n correct. What is
the pr. that he would have done this well if he had no ESP?

Here again we have Bernoulli trials with $p = \frac{1}{2}$. Let
X be a number of correct answers (i.e., proper matches).
Then the required pr. is

$$Pr(X \geq k) \quad = \quad \sum_{j=k}^{n} \binom{n}{j}(1/2)^n.$$

Thus, if getting k or more out of n correctly is an event
of small pr., then this would indicate that the individual
has ESP.

As an example, suppose that n = 100, so the mean and
variance are respectively np = 50, npq = 25, and the
Gaussian approximation is:

$$Pr(X \geq k) \quad = \quad 1 - \Phi((k-50)/5)$$

and the numerical results are: for

$$k = 50 \quad \text{the pr. is} \quad 0.500$$
$$k = 55 \quad \text{the pr. is} \quad 0.136$$
$$k = 60 \quad \text{the pr. is} \quad 0.023.$$

12.4. <u>Telephone calls</u>

Consider a group of n devices which originate calls
(telephone subscribers, lines, etc). The pr. that a device

is busy (i.e., carries a call) is p. In Telephony $p = cT$
where c is the calling rate per unit of time (i.e. the
inverse of the mean time between two consecutive calls
originated by a device) and T is the average duration of
a call; clearly $p < 1$. The pr. that exactly j devices
out of n carry calls is given by binomial $P(j)$ with
parameters n, p. The average number of busy devices np
is called traffic. The pr. that all devices are engaged
is clearly p^n.

Suppose now that these n devices have access to s
lines in some other group, where $s < n$. Thus only s
originated calls can go through, the other being blocked.
We would like to have the pr. of j calls going through,
i.e., not being blocked. This is expressed by the conditional pr.

$$Q(j) = pr(X = j \mid X \leq s), \quad j = 0, 1, \ldots, s.$$

This can be evaluated (as in Section 3) to give

$$Pr(X = j \mid X \leq s) = \frac{pr[(X = j) \cap (X \leq s)]}{pr(X \leq s)} = \frac{pr(X = j)}{pr(X \leq s)}$$

$$= \frac{\binom{n}{j} p^j q^{n-j}}{\sum\limits_{i=0}^{s} \binom{n}{i} p^i q^{n-i}} = \frac{\binom{n}{j} \left(\frac{p}{q}\right)^j}{\sum\limits_{i=0}^{s} \binom{n}{i} \left(\frac{p}{q}\right)^i} \ .$$

The above expression Q is known as the truncated
binomial distribution. Of special interest is that the pr.

of blocking ,defined as the pr. that all s   lines are occupied:

$$Q(s) \;=\; \frac{\binom{n}{s}(\frac{p}{q})^{s}}{\sum\limits_{i=0}^{s}\binom{n}{i}(\frac{p}{q})^{i}} \;.$$

12.5. Cost

Suppose that the cost function φ has the form:

$$\varphi(j) = C > 0, \qquad \text{for } j = 1,\ldots,n-1$$
$$= 0 \qquad\qquad \text{for } j = 0,\, n.$$

Then the average cost is:

$$E\varphi(X) \;=\; C \sum_{j=1}^{n-1} P(j) \;=\; C(1 - P(0) - P(n)) \;=\; C(1 - p^{n} - q^{n}).$$

****

Section 13: Geometric Waiting Time

Consider again Bernoulli trials, but instead of fixing
the number of trials, suppose that we continue trials till
the first success appears. This may happen of course at
the first attempt, or we may perform a large number of trials
-- that is we obtain a succession of failures -- and then
the first success appears. In contrast with the previous
section, the number of trials is now a random variable and
actually the discrete life time. Typical examples are: the

length of a "lucky" string of wins (in a game, in passing examinations), collecting of coupons, time needed to the first occurrence of some event, waiting time, etc.

We consider now the Bernoulli trials with $p = pr(S)$ and $q = pr(F)$, with $p + q = 1$. Let N be a life time of period of F's until the first S, that is the waiting time until the first S. Clearly, N assumes as its value all positive integers: $1, 2, 3, \ldots$. Denote the distribution of N by:

$$P(n) = pr(N = n), \qquad n = 1, 2, \ldots .$$

This is the pr. that the first success occurs exactly at the nth trial. To find $P(n)$ consider events $(N = n)$ for each n, as in the following table:

n	sequence	$P(n)$	(by independence)
1	S	p	
2	FS	pq	
3	FFS	pq^2	
...	...	...	
n	F...FS	pq^{n-1}	(here n-1 F's followed by one S)

Consequently the required distribution is given by:

$$\boxed{P(n) = pq^{n-1}, \qquad n = 1, 2, \ldots}$$.

Observe that $P(n+1)/P(n) = q$, so P decreases with increasing n. The distribution P is called the geometric distribution, because of its relation to the geometric series:

$$\text{for } |\alpha| < 1: \quad \sum_{i=0}^{\infty} \alpha^i = \frac{1}{1-\alpha} .$$

Writing $q = 1 - \alpha$, it follows that:

$$\sum_{n=1}^{\infty} P(n) = 1.$$

The life time N assumes infinitely many values, but the pr. of waiting infinitely long is zero, because $P(n) \to 0$, as $n \to \infty$.

The pr. of waiting more than n is clearly:

$$\text{pr}(N > n) = \sum_{k=n+1}^{\infty} pq^{k-1} = pq^n \sum_{k=n+1}^{\infty} q^{k-n-1} = pq^n/p = q^n.$$

Consequently, the pr. of prolongation of the life time by additional h units is (as in Section 3):

$$\text{pr}(N > n+h \mid N > n) = \text{pr}(N > n+h)/\text{pr}(N > n) = q^{n+h}/q^n = q^h.$$

This pr. does not depend on the duration of life, but only on the amount h of prolongation. Thus, the geometric distribution is memoryless; it is the analogue of the n.e.d. in the continuous case.

The average waiting time $E(N)$ is defined as always by $E(N) = \sum_{n=1}^{\infty} nP(n)$. Direct computation is rather cumbersome: first display the above sum in the following way

$$\sum_{n=1}^{\infty} nP(n) = \begin{cases} p \\ pq + pq \\ pq^2 + pq^2 + pq^2 \\ \cdots \\ pq^n + \ldots + pq^n \quad (n+1 \text{ terms}) \\ \cdots \end{cases}$$

and then sum vertically to get

$$\sum_{k=0}^{\infty} p \left(\sum_{i=k}^{\infty} q^i \right) = p \sum_{k=0}^{\infty} q^k / p = \sum_{k=0}^{\infty} q^k = 1/p.$$

Thus

$$\boxed{E(N) = 1/p} \; ;$$

it can be shown that

$$\boxed{var(N) = qp^{-2}} \; .$$

Note that strictly speaking one should have $0 \leq q < 1$. Indeed, for $q = 0$, obviously

$$P(1) = 1, \quad P(n) = 0 \quad \text{for} \quad n \neq 1, \quad E(N) = 1$$

because then necessarily the first trial results in success. On the other hand, for $q = 1$,

$$P(n) = 0 \quad \text{for all} \quad n, \quad \text{and} \quad pr(N > n) = 1 \quad \text{for all} \quad n,$$

$$E(N) = \infty.$$

One can say that the first success occurs at infinity, that it is never in finite time.

In the intermediary case when $p = q = 1/2$, obviously

$$P(n) = (1/2)^n \quad \text{and} \quad E(N) = 2.$$

Example 1: Suppose that an applicant for a driver's license has 60% chance to pass the test on any given try. Then the pr. that the applicant will eventually get the license on the third try is

$$P(3) = 0.6 \times 0.4^2 = 0.096.$$

Example 2: What is the pr. that in tossing a coin, the first head will appear at the 10th toss? Here $p = 1/2$, so

$$P(10) = (1/2)^{10}$$

Example 3: Let E be the set of all (positive) even integers, and let 0 be the set of all (positive) odd integers. Suppose that positive integers are distributed according to the geometric distribution. Then

$$pr(0) = \sum_{k=0}^{\infty} pq^{2k+1-1} = p \sum_{k=0}^{\infty} q^{2k} = \frac{p}{1-q^2} = \frac{1}{1+q}$$

because for odd $n = 2k + 1$. Therefore,

$$Pr(E) = \frac{q}{1+q} .$$

Note that the sets of even and of odd integers have different pr.'s, despite the intuitive feelings that these sets are "similar." In particular, for $p = q = \frac{1}{2}$, one has

$$pr(O) = \frac{2}{3} \quad \text{and} \quad pr(E) = \frac{1}{3} .$$

Example 4: Assuming the geometric distribution on positive integers, find the pr. that an integer is even when it is divisible by 3. Clearly, this is the conditional pr.:

$$pr(\text{by } 2 \mid \text{by } 3) \;=\; \frac{pr(\text{by } 6)}{pr(\text{by } 3)} .$$

Hence, using the properties of the geometric series:

$$pr(\text{by } 6) \;=\; p \sum_{k=1}^{\infty} q^{6k-1} \;=\; \frac{p}{q} q^6 \sum_{k=1}^{\infty} (q^6)^{k-1} \;=\; \frac{pq^5}{1 - q^6}$$

$$pr(\text{by } 3) \;=\; p \sum_{k=1}^{\infty} q^{3k-1} \;=\; \frac{p}{q} q^3 \sum_{k=1}^{\infty} (q^3)^{k-1} \;=\; \frac{pq^2}{1 - q^3} .$$

Hence the required pr. is:

$$pr(\text{by } 2 \mid \text{by } 3) \;=\; q^3 \frac{1-q^3}{1-q^6} \;=\; \boxed{\frac{q^3}{1+q^3}} .$$

Note that this pr. is always smaller than the pr. that the integer is even (found in Example 3). Indeed:

$$\text{for } q > 0: \quad \frac{q^3}{1+q^3} < \frac{q}{1+q} \;\Rightarrow\; q^3 + q^4 < q + q^4 \;\Rightarrow\; q^2 < 1.$$

Note that for $q = \frac{1}{2}$,   these pr.'s are respectively

$\frac{1}{9} < \frac{1}{3}$.

Section 14: Poisson Distribution

Let us return to Bernoulli trials and to the binomial distribution in Section 11. In many applications the number of trials n is large, and computation of P(j) may be cumbersome. It is therefore of interest to see if some approximation could be found (apart from Gaussian), when n becomes large. Letting n go to infinity would produce a meaningless result, so this limit must be taken with some restriction to obtain a reasonable answer. Indeed, in many cases one encounters the situation that when the number of trials n **becomes** very large, at the same time the pr. of success p becomes very small (the so called rare events), yet the average np remains practically the same. This suggests checking the following limit:

$\lim_{n\to\infty} P(j)$ **with p → 0 in such a way that**

np = μ is constant.

This turns out to be the right approach. To compute the limit the use must to made of the Stirling approximation to the factorial:

$$n! \approx \sqrt{2\pi}\, n^{n} e^{-n}$$

and the following limit from calculus:

$$\lim_{n\to\infty}(1 - \frac{a}{n})^n = e^{-a}, \quad \text{for any } a.$$

Now we can write:

$$\binom{n}{j} p^j(1-p)^{n-j} = \frac{n!}{j!(n-j)!} p^j(1-p)^{n-j}$$

$$= \frac{n^n e^{-n}}{j!(n-j)^{n-j} e^{-n+j}} (\frac{\mu}{n})^j (1 - \frac{\mu}{n})^{n-j}$$

$$= \frac{\mu^j}{j!} (\frac{n}{n-j})^{n-j} (1 - \frac{\mu}{n})^{n-j} e^{-j} \to \frac{\mu^j}{j!} e^{-\mu}.$$

This yields the result:

$$\boxed{P(j) = \frac{\mu^j}{j!} e^{-\mu}, \quad j = 0, 1, \ldots}.$$

This is known as the <u>Poisson</u> <u>distribution</u>. Observe that j ranges from 0 to infinity (as n is now infinite). $P(j)$ gives the pr. of exactly j successes in the infinite number of trials. Here the Poisson distribution has been obtained as the approximation to the binomial. We shall see later that it stands on its own merits; and can be derived independently of the binomial.

We must first verify that

$$\sum_{j=0}^{\infty} P(j) = 1.$$

This follows immediately from the series representation for e:

$$\sum_{j=0}^{\infty} a^j/j! = e^a.$$

From $P(j+1)/P(j) = \mu/(j+1)$ it is easy to see that $P(j)$ increases and then decreases with j, when $\mu > 1$, the maximum occurring for j being the integer between $\mu - 1$ and μ; for $\mu \le 1$, $P(j)$ decreases with j.

It is very easy to find the mean of a random variable with the Poisson distribution:

$$E(X) = \sum_{j=0}^{\infty} jP(j) = \sum_{j=1}^{\infty} j \frac{\mu^j}{j!} e^{-\mu}$$

$$= \mu e^{-\mu} \sum_{j=1}^{\infty} \mu^{j-1}/(j-1)! = \mu e^{-\mu} e^{\mu} = \mu.$$

To get variance, consider first $EX(X-1)$; by the same argument

$$EX(X-1) = \sum_{j=2}^{\infty} j(j-1) \frac{\mu^j}{j!} e^{-\mu} = \mu^2 e^{-\mu} \sum_{j=2}^{\infty} \mu^{j-2}/(j-2)!$$

$$= \mu^2 e^{-\mu} e^{\mu} = \mu^2.$$

Hence $EX^2 = \mu^2 + \mu$, so $\text{var } X = EX^2 - (EX)^2 = \mu^2 + \mu - \mu^2 = \mu$. Thus, we have a very interesting property of the Poisson distribution:

$$\boxed{EX = \text{var } X = \mu} \quad ,$$

where μ is the parameter occurring in the formula $P(j)$.

When it is required to find the distribution function of X, denoted F(k) with k restricted to non-negative integers, one has simply to add:

$$F(k) = \text{pr}(X \le k) = \sum_{j=0}^{k} P(j) = \sum_{j=0}^{k} \frac{\mu^j}{j!} e^{-\mu}.$$

The values of P(j) and F(k) are tabulated for given μ.

As an exercise in calculus, we can find an interesting expression for F(k) by treating F as a function of μ (for constant k). Differentiation with respect to μ yields

$$\frac{dF}{d\mu} = \sum_{j=1}^{k} \frac{\mu^{j-1}}{(j-1)!} e^{-\mu} - \sum_{j=0}^{k} \frac{\mu^j}{j!} e^{-\mu} = -\frac{\mu^k}{k!} e^{-\mu}.$$

Then, integration, with initial condition F = 1 for $\mu = 0$, yields

$$F(k) = 1 - \frac{1}{k!} \int_0^{\mu} t^k e^{-t} dt = \frac{1}{k!} \int_{\mu}^{\infty} t^k e^{-t} dt.$$

****

Example 1: A book of 200 pages contains 100 misprints. Find the pr. that a given page contains at least 2 misprints. The average number of misprints per page is $\mu = 100/200 = 0.5$, so

$$\text{pr}(X \ge 2) = 1 - F(1) = 1 - 0.9098 = 0.0902.$$

Example 2: The average number of calls received by a switchboard during a fixed period is 5. What is the pr. that 6

or less calls will be received? Clearly $F(6) = 0.7622$
(for $\mu = 5$).

Example 3: Let E be the set of all (nonnegative) even
integers, and let 0 be the set of all (nonnegative) odd
integers. Suppose that nonnegative integers are distributed
according to the Poisson distribution. Then

$$\text{pr}(E) = \sum_{i=0}^{\infty} \frac{\mu^{2i}}{(2i)!}\, e^{-\mu} = (1 + \frac{\mu^2}{2!} + \frac{\mu^4}{4!} + \ldots)e^{-\mu}$$

$$= e^{-\mu} \cosh \mu,$$

using the series expansion for cosh, and the relation
$\cosh \mu + \sinh \mu = e^{\mu}$. Hence:

$$\text{pr}(0) = e^{-\mu} \sinh \mu.$$

Example 4: Colorblindness appears in 1% of people in a
certain population. How large must a random sample (with
replacement) be if the pr. of its containing a colorblind
person is to be 0.95 or more.

The pr. of at least one is $1 - P(0) = 1 - e^{-\mu} \geq 0.95$
but $\mu = np = n/1000$, so

$$e^{-\mu} \leq 0.05, \quad \text{or} \quad n \geq 300.$$

Example 5: What is the pr. that in a class of 110 students
exactly 2 will have birthdays today? These are Bernoulli

trials with $p = 1/365$ and $n = 110$, so $\mu \approx 0.3$, and $P(2) = 0.033$.

Example 6: Suppose that only s lines out of a (infinite) group can be connected to the second stage in a telephone system. Find the pr. that exactly j lines are occupied, given that this number does not exceed s.

This is clearly

$$pr(X = j \mid X \leq s) \ = \ \frac{pr(X = j)}{pr(X \leq s)} \ = \ \frac{P(j)}{F(s)} \ = \ \frac{\frac{\mu^j}{j!}}{\sum_{i=0}^{s} \frac{\mu^i}{i!}}$$

for $j = 0,1,\ldots,s$.

This is known as the Erlang distribution (or truncated Poisson): its value for $j = s$ is the famous Erlang formula for pr. of blocking.

Section 15: Accidents Just Happen

In consideration of Bernoulli trials (Section 11) the number of trials was fixed, but the number of successes was a random variable. In the discussion of the waiting time (Section 13), the number of trials till the first success was the random variable. In many problems, however, both the number of trials and the number of successes are random variables. We shall now consider such a situation, and we will be interested in finding the distribution of successes.

Consider a fluctuating population of some objects of a specified kind (say bacteria in a culture, accidents in a city, cars in some region, houses in a community, eggs in a shopping basket) and assume that the number of objects N has Poisson distribution with mean λ. That is, the $\mathrm{pr}(N = n)$ $= Q(n)$ of having exactly n objects is given by

$$Q(n) = \frac{\lambda^n}{n!} e^{-\lambda}, \qquad n = 0,1,\dots .$$

Suppose that these objects may suffer some misfortune like damage, loss etc., which for convenience we shall call an accident. (For example, death of a bacteria, fatal accident in a city, broken car, house fire, rotten egg). Assume that accidents occur independently, each with pr. p. That is, we have here Bernoulli trials, and the pr. of exactly j accidents among n objects, denoted by $P(j|n)$, is now

$$P(j|n) = \binom{n}{j} p^j (1-p)^{n-j}, \qquad j = 0,1,\dots,n.$$

We will be interested in the pr. that there will be exactly j accidents, irrespective of the number of objects in the population. Denote by S the number of accidents; we wish to find

$$\mathrm{pr}(S = j) = P(j), \qquad j = 0,1,\dots .$$

Clearly, we are looking for the distribution of dead bacteria,

fatal accidents, broken cars, fires, rotten eggs, as the case
may be.

From the above description it is clear that we have
"successes" (i.e., accidents) in n Bernoulli trials, so in
fact the conditional pr.:

$$\text{pr}(S = j \mid N = n) \;=\; P(j \mid n).$$

By properties of conditional pr. (see Section 3), the joint
pr. is

$$\text{pr}(S = j,\, N = n) \;=\; P(j \mid n)Q(n).$$

Taking summation over all n, we obtain the marginal distribution:

$$\text{pr}(S = j) \;=\; \sum_{n} P(j \mid n)Q(n)\;.$$

This is the required distribution $P(j)$.

It should be emphasized that here the binomial distribution $P(j \mid n)$ and the Poisson distribution $Q(n)$ are quite
separate and not related to each other. (That's why mean
has been denoted by λ in order to avoid confusion with
$\mu = np$ from Section 14).

The explicit calculation of $P(j)$ proceed easily as
follows: Observe first that summation in the above expression is actually from $n = j$, becuase it is impossible to

have $n < j$. Note that this is automatically taken care of by the fact that

$$\binom{n}{j} = 0 \quad \text{for} \quad n < j.$$

Hence

$$\boxed{P(j)} = \sum_{n=j}^{\infty} \binom{n}{j} p^j (1-p)^{n-j} \frac{\lambda^n}{n!} e^{-\lambda} = \frac{p^j \lambda^j}{j!} e^{-\lambda} \sum_{n=j}^{\infty} \frac{(1-p)^{n-j} \lambda^{n-j}}{(n-j)!}$$

$$= \frac{p^j \lambda^j}{j!} e^{-\lambda} e^{(1-p)\lambda} = \boxed{\frac{(\lambda p)^j}{j!} e^{-\lambda p}} \quad , \quad j = 0, 1, \ldots .$$

Thus we obtain the very interesting result that the number of accidents is also Poisson distributed, but with mean λp. So

$$E(S) = \lambda p = \text{var } S.$$

The effect of individual accidents is reflected in appearance of factor p, which reduces the mean λ to the new mean λp. Recall that if Z is a r.v. equal to one if a success occurs, and to zero if not, then $EZ = p$. Hence we can write

$$ES = EN \cdot EZ$$

which has an intuitive meaning: the average number of accidents equals the average number of objects times the average value of accident proneness.

$$* * * * * * * * * * * * * *$$

As another illustration, suppose that the population distribution $Q(n)$ is geometric. As in the present situation n ranges from zero, the geometric distribution is taken of the form:

$$Q(n) = (1-a)a^n, \quad n = 0,1,\ldots, \quad 0 < a < 1.$$

Here $EN = a(1-a)^{-1}$.

Hence, as before:

$$
\begin{aligned}
P(j) &= \sum_{n=j}^{\infty} \binom{n}{j} p^j (1-p)^{n-j} (1-a) a^n \\
&= (1-a) a^j p^j \sum_{n=j}^{\infty} \binom{n}{j} [a(1-p)]^{n-j} \\
&= (1-a)(ap)^j \sum_{k=0}^{\infty} \binom{j+k}{j} [a(1-p)]^k \\
&= \frac{(1-a)(ap)^j}{[1-a(1-p)]^{j+1}} = \frac{1-a}{1-a+ap} \left(\frac{ap}{1-a+ap}\right)^j
\end{aligned}
$$

where in the last step the use has been made of a relation

$$\sum_{k=0}^{\infty} \binom{j+k}{j} a^k = \frac{1}{(1-a)^{j+1}}.$$

Thus, $P(j) = (1-b)b^j$, with $b = ap(1-a+ap)^{-1}$; the result being the geometric distribution again, but with mean

$$ES = ap(1-a)^{-1}.$$

Note that the relation $ES = p \cdot EN$ holds again.

$$**************$$

This result concerning the mean is not, however, an accident but is true for our Bernoulli trials no matter what the distribution $Q(n)$ is. Indeed, for the average number of accidents we have:

$$E(S) = \sum_j jP(j) = \sum_n \sum_j jP(j|n)Q(n) = p\sum_n nQ(n) = pE(N)$$

because for the binomial distribution

$$\sum_j jP(j|n) = np.$$

Remark: It is of great interest that our argument can be extended to more general situations. This will become clear when we shall take another look from a different angle at the problem on hand.

Consider independent identically distributed r.v.'s $Z_1, \ldots, Z_n$ with common mean $E(Z)$. Write for their sum (as in Section 5):

$$S_n = Z_1 + \ldots + Z_n.$$

This is for fixed n. When however the number of terms is a random variable itself, say N, then we get the so called random sum:

$$S = Z_1 + \ldots + Z_N$$

(also denoted by S_N, and called a compound r.v.). Then, we have from properties of conditioning:

$$pr(S=j) = \sum_n pr(S=j, N=n) = \sum_n pr(S=j \mid N=n)pr(N=n).$$

Assuming independence of N from all Z_i's, we have:

$$P(j \mid n) \ = \ pr(S = j \mid N = n) \ = \ pr(S_n = j).$$

Hence

$$pr(S = j) \ = \ \sum_n pr(S_n = j) pr(N = n).$$

In the special case when Z_i are Bernoulli r.v.'s (as in sub-section 11.3), we obtained the situation of this section, and the formula for $pr(S = j)$ is the same as that stated at the beginning of the section.

In the general case considered now, $P(j \mid n)$ is not of the binomial form, but nevertheless we have in view of the above new look:

$$\sum_j jP(j \mid n) \ = \ E(S_n) \ = \ nE(Z)$$

so finally

$$E(S) \ = \ \sum_n nQ(n) \ = \ E(Z)E(N).$$

Indeed, this relation for expectation is what we would expect intuitively!

Chapter 2: Problems

1. At a small art exhibition there are six paintings includ-
    ing one by van Gogh. It is expected that on the average
    2 buyers will express sufficient interest to ask for
    the price of the van Gogh painting. What is the proba-
    bility of this happening?

2. A superstitious commuter driving to work must pass n
    traffic lights, and assumes that a green light shows with
    pr. 1/2.

    The commuter believes that the day will be lucky if
    the number of green lights encountered is even, and then
    assigns the value of +1 to such a day. If the number
    of green lights is odd, the day is considered to be bad,
    and the value of -1 is assigned to it.

    Find the average value assigned to a day (assuming
    binomial distribution.).

3. Suppose that the number of busy machines in a laundromat
    with n washing machines, follows the binomial distri-
    bution with p being the pr. that a machine is busy.

    According to the management, profit is proportional
    to the number of operating machines, except in the situ-
    ation when all n machines are working; the profit is
    then zero (because of high energy consumption,
    dissatisfied waiting customers, etc.).

    Thus, the profit function is

    $$\varphi(j) = \begin{cases} cj & \text{for } 0 \le j \le n-1 \\ 0 & \text{for } j = n \end{cases}, \quad (c > 0).$$

Show that:

(i) The average profit is

$$E\varphi(X) = np(1-p^{n-1})c.$$

(ii) The average profit has maximum when

$$p = (1/n)^{\frac{1}{n-1}}, \qquad \text{for } n > 1.$$

4. Let N be a discrete life time with binomial distribution (with parameters n and p). Suppose that the cost function φ has the form

$$\varphi(j) = z^j \qquad \text{for } j = 0, 1, \ldots, n, \qquad |z| \leq 1.$$

Show that the average cost is given by

$$E\varphi(N) = (pz + 1 - p)^n.$$

Sketch the graph of the average cost as a function of z, and verify that its slope at $z = 1$ is equal to the mean np.

5. Using Gaussian approximation, compute the pr. that the number of successes in 100 Bernoulli trials, with mean 80 differs in absolute value from the mean by 8 or less:

$$pr(|X-80| \leq 8).$$

Compare with the Chebyshev inequality.

6. Let X be a discrete r.v. with binomial distribution (with parameters n and p).

(i)   Find the second moment $\mathbb{E}(X^2)$, and plot its
graph as a function of p (for fixed n). Verify
that always $\mathbb{E}(X^2) \leq n^2$.

(ii) Find p such that the second moment equals
$k[\mathbb{E}(X)]^2$, where k is a positive integer.

(iii) Find p which maximizes $\mathrm{var}(X)$, for fixed n.

7.   (i)   Sketch the graph of the binomial distribution
$P(j)$, for fixed j, as a function of p. Consider separately three cases: $j = 0$, $1 \leq j \leq n-1$,
$j = n$.

(ii) Consider for binomial distribution:

$$F(j) = \mathrm{pr}(X \leq j) = \sum_{i=0}^{j} P(i)$$

defined as in the text (Section 11).
Show that for each j

$$F(j) = n\binom{n-1}{j} \int_{p}^{1} t^j (1-t)^{n-1-j}\, dt.$$

Hint: differentiate both sides with respect to
p, and use properties of binomial coefficients.

(iii) Sketch the graph of $F(j)$, for fixed j, as a
function of p. What are the values for $p = 0$,
$p = 1$?

8. Suppose that a discrete life time N has distribution
    given by:

$$p(n) = \frac{\lambda^{n-1}}{(1+\lambda)^n} \quad \text{for} \quad n = 1,2,\ldots$$

where $\lambda > 0$ is a constant.

(i) Show that $EN = \lambda + 1$, var $N = \lambda(\lambda+1)$.

(ii) Take $\lambda = 16/9$ and use the Gaussian approximation to
compute the probability that $N \leq 5$.

9. A repairman never shows up on the day he is called for
    service, but a customer must wait for one, two or more
    days. Suppose that the average waiting time is 3 days.
    Find the probability of the waiting time being less than
    the average (assume a geometric distribution).

10. Suppose that the average waiting time for the first suc-
    cess in Bernoulli trials equals $3\frac{1}{2}$.
    Show that the pr. of waiting more than n (units
    of time) for the first success is $(5/7)^n$.

11. A commuter driving home from work observes a "lucky"
    sequence of green traffic lights (assume the geometric
    distribution with $p = 1/2$). Find the pr. that

(i) The first red (i.e. not green) light occurs at
the n-th traffic light intersection;

(ii)   The first red light occurs sometime later after passing successfully n lights;

(iii) The first red light occurs after h additional traffic lights, having passed successfully n lights.

Compute these pr.'s for n = 1,2,3,4, and h = 1.

12. For the geometric distribution write

$$1 = \sum_{n=1}^{\infty} p(1-p)^{n-1}.$$

Differentiate both sides with respect to p and deduce that

$$\sum_{n=1}^{\infty} np(1-p)^{n-1} = 1/p.$$

13. Regard the geometric distribution

$$P(n) = p(1-p)^{n-1}$$

as a function of p, for fixed n: Show that

(i) $dP/dp = (1-np)(1-p)^{n-2}$ for n = 1,2,... ;

(ii) $d^2P/dp^2 = (n-1)(np-2)(1-p)^{n-3}$ for n = 1,2,...;

(iii) P has a maximum at p = 1/n for n = 2,3,...;

(iv) P has an inflection point at $p = 2/n$ for
$n = 3,4,\ldots;$

(v) plot the graphs of P as a function of p for
$n = 1$, $n = 2$ and $n = 3$. Indicate the slope
at $p = 0$ and $p = 1$.

14. Let N be a discrete life time with geometric distribution. Suppose that the cost function φ is of the
form:

$$\varphi(n) = z^n \quad \text{where} \quad |z| \le 1, \quad n = 1,2,\ldots .$$

(i) Show that the average cost, regarded as a function of z, is

$$E\varphi(N) \equiv K(z) = pz(1-qz)^{-1}.$$

(ii) Deduce that

$$dK/dz = p(1-qz)^{-2}, \qquad d^2K/dz^2 = 2pq(1-qz)^{-3}.$$

(iii) Verify that

$$EN = dK/dz\Big|_{z=1} = 1/p,$$

$$EN(N-1) = d^2K/dz^2\Big|_{z=1} = 2q/p^2$$

and deduce that

$$\text{var } N = q/p^2.$$

15.   Compute for the geometric distribution:

$$pr(|N-EN| > \varepsilon), \quad \text{for} \quad \varepsilon > 0,$$

and compare it with the Chebyshev approximation.

16.   Suppose that on the average 5 of your friends eat
      lunch daily at the Student Union. Using the Poisson
      distribution, calculate the pr. that on a particular
      day you will meet at lunch:

      (i)   no friend at all;

      (ii)   exactly one friend;

      (iii) more than one friend.

      Do these figures agree with your experience?

17.   In Criminology, it is found that the number X of
      people having finger prints belonging to a certain type
      is distributed according to the Poisson distribution
      with a parameter μ (representing the average number of
      people having the same type). For identification pur-
      poses (by detectives, in courts, etc.) it is of great
      interest to know the pr. of duplication, i.e., the pr.
      that some other person has finger prints of the same
      type as the accused.

         Formally, this is defined as the pr. of two or more
      people of the same type, when it is known that at least
      one person has this type, i.e. $\mathbb{P}(X \geq 2 \mid X \geq 1)$.

Show that the pr. of no duplication is given by

$$\mathbb{P}(X = 1 \mid X \geq 1) \;=\; \frac{\mu}{e^{\mu}-1} \;.$$

Calculate this pr. for $\mu = 1/2$, $\mu = 1$, $\mu = 5$. Does this pr. decrease with increasing μ ? Try to sketch the graph .of this pr. as a function of μ.

18. Suppose that the pr. of an individual having an infection is $p = 1/25$. Compute the pr. that in a group of 100 people more than 6 people will have an infection, using the

   (i) Poisson approximation;

   (ii) Gaussian approximation.

19. If there are on the average 1% left handers, find the pr. of having at least 5 left handers amoung 300 people.

20. Estimate the number of raisins which a cookie should contain on the average if it is desired that the pr. of a cookie to contain at least one raisin to be 0.99 or more.

21. Let X be a discrete r.v. with Poisson distribution
     (with parameter μ).

     (i) Show that

$$\text{var}(X/\sqrt{\mu}) = 1.$$

     (ii) Let $Y = (X - \mu)^2$. Show that

$$EY^2 \geq \mu^2.$$

22. Regard the Poisson distribution

$$P(j), \qquad j = 0,1,\ldots, \quad \mu > 0$$

     as a function of its mean μ, for fixed j. Show that:

     (i) $dP/d\mu = P(j-1)(1-\mu/j)$ for $j \neq 0$, and

          $dP/d\mu = -P(0)$ for $j = 0$;

     (ii) $d^2P/d\mu^2 = P(j-2)[1 - \frac{2j\mu-\mu^2}{j(j-1)}]$ for $j = 2,3,\ldots$

               $= P(0)(\mu-2)$ for $j = 1$

               $= P(0)$ for $j = 0$.

     (iii) P has maximum for $j = 1,2,\ldots$ at $\mu = j$

     (iv) P has two inflection points at $\mu = j \pm \sqrt{j}$, for
           $j = 2,3,\ldots$, and one inflection point at $\mu = 2$
           for $j = 1$.

     (v) Plot graphs of P as a function of μ for
          $j = 0,1,2$ and 3.

23. Show that for the Poisson distribution $F(j) = \sum_{i=0}^{j} P(i)$ can be written as

$$F(j) = \frac{1}{j!} \int_{\mu}^{\infty} t^j e^{-t} \, dt$$

(see Section 14). Show that $dF(j)/d\mu = -P(j)$.

24. Write the Erlang formula for blocking in the form

$$B = P(s)/F(s)$$

(see Example 6, Section 14), and deduce that

$$dB/d\mu = B[B + (s-\mu)/\mu].$$

25. Let N be a discrete life time with Poisson distribution. Suppose that the cost function φ is of the form:

$$\varphi(j) = z^j \quad \text{where} \quad |z| \leq 1, \quad j = 0,1,\dots .$$

   (i) Show that the average cost, regarded as a function of z, is

$$E\varphi(N) \equiv K(Z) = e^{-\mu(1-z)}.$$

   (ii) Deduce that

$$dK/dz = \mu K(z), \qquad d^2K/dz^2 = \mu^2 K(z).$$

   (iii) Verify that

$$EN = dK/dz \Big|_{z=1} = \mu,$$

$$EN(N-1) = d^2K/dz^2 \Big|_{z=1} = \mu^2$$

and deduce that

$$var\ N = \mu.$$

26. Suppose that the life time N has Poisson distribution with mean μ. Let the cost function be $(-1)^n$ for $n = 0,1,\dots$. Show that the average cost is $e^{-2\mu}$. (Note: this is immediate! Contrast it with Example 3 in Section 14).

27. Compute for the Poisson distribution

$$pr(|N-EN| > \varepsilon),\quad for\ \varepsilon > 0,$$

and compare it with the Chebyshev approximation.

28. Let X be a discrete life time assuming values $0,1,2,\dots,K$ (finite or infinite). Suppose that the cost function φ is of the form:

$$\varphi(j) = z^j \quad for\ |z| \le 1.$$

Then, the average cost, regarded as a function of z, and defined by

$$C(z) = E(z^X)$$

has properties analogous to the Laplace transforms (see Section 7):

(i) $C(z)\Big|_{z=1} = 1, \quad dC(z)/dz\Big|_{z=1} = E(X),$

$d^2C(z)/dz^2\Big|_{z=1} = EX(X-1).$

(ii) Let $X_1, \ldots, X_n$ be independent identically distributed life times with common distribution with average cost $C(z)$. Consider the total life time $S_n = X_1 + \ldots + X_n$. Show that the average cost for S_n is

$$E(z^{S_n}) = [C(z)]^n$$

(this is the analogue of convolutions in Section 5).

(iii) Show that if X assumes values 0 and 1 only, with pr.'s q and p, respectively, then S_n has the binomial distribution with parameters n, p. (This is the same derivation as that in subsection 11.3, see also Problem 4.)

(iv) Show that if $X_1, \ldots, X_n$ have common binomial distribution with parameters k, p, then S_n has also a binomial distribution but with parameters nk, p, (see Problem 4).

(v) Show that if $X_1, \ldots, X_n$ have common Poisson distribution with mean μ, then S_n has also Pois-

son distribution but with mean $n\mu$ (see Problem
25).

(vi)   Show that if $X_1,\ldots,X_n$ have common geometric
       distribution pq^{j-1} $(j = 1,2,\ldots)$, then S_n
       has so called "negative binomial distribution"
       of the form

$$P(j) = \binom{j-1}{j-n}p^n q^{j-n}, \quad \text{for } j \geq n$$

(see Problem 14).

29. With reference to discrete life times $Z_1,\ldots,Z_n$ and
     S and N in Section 15, consider the following aver-
     age costs:

$$C(z) = E(z^Z), \qquad U(z) = E(z^N) = \sum_{n=0}^{\infty} z^n Q(n),$$

$$W(z) = E(z^S) = \sum_{j=0}^{\infty} z^j P(j).$$

(i)   Using the formula for $P(j)$, show that

$$W(z) = U[C(z)].$$

(ii) Deduce that

$$E(S) = E(Z)E(N)$$

$$\text{var } S = (\text{var } Z)E(N) + (\text{var } N)(EZ)^2.$$

(iii) Using the above relations, check var S in the
      examples worked out in Section 15.

30. Suppose that in Problem 29: $C(z) = 1 - \sqrt{1-z}$ and that

Q(n) is the Poisson distribution with mean λ. Show

that

$$W(z) = e^{-\lambda\sqrt{1-z}} .$$

What about E(S) ?

Chapter 3

To Renew or not to Renew

Renewal processes are very intuitive and rather easy to handle (at least at the beginning--in later stages things start to be awkward), and have many practical applications. Indeed, we shall talk now about situations familiar from everyday life -- replacement or renewal of worn out appliances, devices etc., and other successive life times.

Although the mathematical aspects may be more complex than what we have seen so far, you should have no difficulty in following the presentation, if treated with compassion. Renewals are described in Section 16, and Section 17 presents the glorious achievements of the theory. One may often wonder why so much work is needed to justify so "obvious" results. As a consolation, we may be happy to see that our mathematics agrees with our intuition.

The story of the rabbit (Section 18) as sad as it may be, should stop us for awhile to reflect on what lies in the background. We only remark here that we touched upon the extension of renewals to random walks, and on the modern

161

concept of stopping times (a vast treasure for the mathematically minded!).

Section 16: Renewals

Consider the following homely situation. A light bulb
is shining brightly in your room, and suddenly it blows up.
You replace it by another one -- it lasts for a duration of
its life and again burns itself out. You replace it again,
and again it blows up, so again... . Two obvious questions
present themselves:

1) What is the distribution of the total time needed to use
    all your supply of bulbs?

2) What is the distribution of the total number of bulbs
    during a year?

In the first question the r.v. is the total life time, in
the second question the r.v. is the number of bulbs; the
first one is the continuous r.v., whereas the second is discrete. We have been discussing such r.v.'s separately, and
now it is time to consider them jointly. It is clear that
both questions are related to each other. It may be more
convenient for the reason of generality to speak about
renewals, that is immediate replacements of objects (bulbs,
cars, machines, persons, etc.) at the instant of termination
of a life time. Thus we would like to have a distribution
of

1) total life time for n renewals,

2) total number of renewals within time t.

Denote by X_n the life time of the object immediately pre-
ceeding the n-th renewal (life time of the n-th bulb),
where $n = 1,2,...$. We shall assume that all life times
X_n are independent, identically distributed (i.i.d.) r.v.'s
with a common d.f. F having density f, and with mean μ
and variance σ^2. That is we are in the situation in Section 2:

$$pr(X_n \leq t) = F(t) \quad \text{for} \quad 0 \leq t < \infty \quad \text{(independent of n).}$$

Total time for n renewals is clearly

$$S_n = X_1 + ... + X_n.$$

Write for its d.f.:

$$pr(S_n \leq t) = G_n(t), \quad 0 \leq t < \infty, \quad n = 1,2,... .$$

Clearly, S_n are nonnegative, and $G_1 = F$. We already
have noted that $E(S_n) = n\mu$, and (by independence) $var(S_n)$
$= n\sigma^2$; see Section 5.

Our first problem will be solved if we can find the
distribution of S_n. We have done it already in Section 5.
Consider first $n = 2$, so $S_2 = X_1 + X_2$. As is shown in Sec-

tion 5, S_2 has density g_2 given by the convolution

$$g_2(t) = \int_0^t f(t-s)f(s)\,ds.$$

Now, write $S_n = S_{n-1} + X_n$; r.v.'s S_{n-1} and X_n are independent, and we can apply to their densities the same argument, to get

$$g_n(t) = \int_0^t f(t-s)g_{n-1}(s)\,ds, \qquad n = 2,3,\ldots, \quad 0 \le t < \infty.$$

Thus, starting with $g_1 = f$, we may compute all g_n from the above recurrence relation. Unfortunately, actual calculations are usually complicated.

It may be convenient to define $S_0 = 0$, so its d.f. $G_0 = 1$ for all t. Then, remembering that G_n is the integral of g_n (for $n = 1,2,\ldots$), we can write the above recurrence relation in the form

$$\boxed{G_n(t) = \int_0^t F(t-s)\,dG_{n-1}(s), \qquad 0 \le t < \infty, \quad n = 1,2,\ldots}$$

Recall that G_n is the pr. that the total life time of n renewals is t or less. Moreover, it has been assumed that the 0-th renewal (i.e. the beginning of the first life time) is at the origin. This provides the solution for question 1.

The collection of r.v.'s (S_n) is called a <u>renewal</u> <u>process</u>.

<u>Remark</u>: X_n represents a life time of the n-th item (installed in a certain system), that is the time interval between the (n-1)-th and the n-th renewals. S_n is then the total time for n renewals, or the time at which the n-th renewal takes place.

Although the G_n are hard to compute, they are helpful to solve Problem 2. Recall that we wish to find the distribution of a number of renewals in a fixed time interval from 0 to t (excluding the instant 0, but including the instant t). More precisely, define for each t a r.v. N_t representing the number of renewals that will have occurred by the time t, including any made at t but excluding the initial (the 0-th). Write for the distribution of N_t:

$$pr(N_t = n) = P_t(n), \quad n = 0, 1, \ldots, \quad t > 0.$$

Observe that $P_t(n)$ is the distribution in n; it represents the number of renewals that may take place. In particular, for n = 0, $P_t(0)$ is the pr. that no renewal took place; that is the original item is still in progress at time t. Clearly, the r.v. N_t is discrete. It should be stressed that t is a parameter. Thus, we have a family of r.v.'s (N_t) -- such a family is known as a stochastic process -- (N_t) is also called a <u>renewal</u> <u>process</u>. And for each t, the r.v. N_t has its distribution P_t.

To find P_t, note that the value of N_t gives the index n for which the next sum S_{n+1} overshoots the point

t for the first time. Hence, the event of exactly n
renewals up to the time t is:

$$(N_t = n) \quad = \quad (S_n \leq t < S_{n+1}) \quad = \quad (S_n \leq t) - (S_{n+1} \leq t).$$

Taking pr.'s of both sides, one finds that

$$\boxed{P_t(n) \quad = \quad G_n(t) - G_{n+1}(t), \quad n = 0,1,2,\ldots, \quad t > 0}.$$

This is the basic relation, solving Problem 2. Note that
for n = 0 we have obviously:

$$P_t(0) = 1 - F(t).$$

It is of interest to note that in view of the convolution
expression for G_{n+1} stated above, we can write:

$$P_t(n) \quad = \quad \int_0^t [1-F(t-s)] \, dG_n(s), \quad n = 1,2,\ldots \; .$$

This relation is useful on many occasions.

Another relation between P_t and G_n is obtained as
follows. Write $P_t(n)$ for each n:

$$P_t(0) \quad = \quad 1 \quad - G_1(t)$$

$$P_t(1) \quad = \quad G_1(t) \quad - G_2(t)$$

$$\vdots \qquad\qquad \vdots \qquad\qquad \vdots$$

$$P_t(n-1) \quad = \quad G_{n-1}(t) - G_n(t).$$

Adding, one has

$$\sum_{k=0}^{n-1} P_t(k) = 1 - G_n(t), \qquad n = 1,2,\ldots$$

which is nothing else but

$$\text{pr}(N_t \leq n-1) = \text{pr}(S_n > t)$$

a useful relation which expresses duality between N_t and S_n.

It can be verified that

$$G_n(t) = \sum_{k=n}^{\infty} P_t(k), \qquad n = 0,1,2,\ldots .$$

In particular, for $n = 0$, this shows that for each $t > 0$:

$$\sum_{k=0}^{\infty} P_t(k) = 1.$$

It is clear from this that $\lim_{n \to \infty} G_n(t) = G_\infty(t) = 0$ identically, so S_n increases to ∞ with pr. one.

Example: Suppose that the common lifetime is exponential

$$F(t) = 1 - e^{-\lambda t}, \quad t \geq 0, \quad \lambda > 0,$$

with density

$$f(t) = \lambda e^{-\lambda t}, \qquad t > 0, \quad \lambda > 0.$$

As shown in Section 5, $S_2 = X_1 + X_2$ has density

$$g_2(t) = \lambda t e^{-\lambda t}, \qquad t > 0.$$

Try once more (for $n = 3$):

$$g_3(t) = \int_0^t \lambda s e^{-\lambda s}\lambda \cdot \lambda e^{-\lambda(t-s)} \, ds = \frac{(\lambda t)^2}{2} e^{-\lambda t}\lambda.$$

It can be verified (by induction) that the g_n have the general form:

$$g_n(t) = \frac{(\lambda t)^{n-1}}{(n-1)!} e^{-\lambda t}\lambda, \qquad t > 0.$$

Hence, the d.f. G_n is

$$G_n(t) = \int_0^t \frac{(\lambda\tau)^{n-1}}{(n-1)!} e^{-\lambda\tau}\lambda d\tau$$

(by substitution $\lambda\tau = x$)

$$= \frac{1}{(n-1)!} \int_0^{\lambda t} x^{n-1} e^{-x} \, dx$$

integrating by parts, or using the integral from Section 14):

$$= 1 - \sum_{k=0}^{n-1} \frac{(\lambda t)^k}{k!} e^{-\lambda t}.$$

Hence,

$$P_t(n) = G_n(t) - G_{n+1}(t) = \sum_{k=0}^{n} \frac{(\lambda t)^k}{k!} e^{-\lambda t} - \sum_{k=0}^{n-1} \frac{(\lambda t)^k}{k!} e^{-\lambda t}$$

$$= \frac{(\lambda t)^n}{n!} e^{-\lambda t}, \qquad n = 0,1,2,\ldots, \qquad t > 0.$$

Thus, we get the Poisson distribution!! This derivation of the Poisson distribution is much more important than that from Section 14. It stresses the importance of the Poisson distribution on its own merits.

Note: $P_t(0) = e^{-\lambda t}$.

Section 17: Renewal Equation

In the previous section we found the distribution P_t of the number N_t of renewals up to time t (including t). We also have seen that the calculations may be cumbersome. Thus, it will be of more interest to find the average number of renewals up to t, that is $E(N_t)$. It is a pleasant surprise that this expected value can be calculated directly, without explicit knowledge of the distribution P_t.

Write for the renewal function:

$$U(t) = E(N_t)$$

Then using results from the previous section:

$$U(t) = \sum_{n=0}^{\infty} nP_t(n) = \sum_{n=1}^{\infty} n[G_n(t) - G_{n+1}(t)]$$

$$= G_1(t) - G_2(t) + 2[G_2(t) - G_3(t)] + 3[G_3(t) - G_4(t)] + \ldots$$

$$= G_1(t) + G_2(t) + G_3(t) + \ldots$$

$$= \sum_{n=1}^{\infty} G_n(t).$$

This is a very interesting formula; the drawback is that we need all G_n. However, we can do much better than that! Proceed as follows:

$$U(t) = G_1(t) + \sum_{n=2}^{\infty} G_n(t) \qquad \text{splitting the sum}$$

$$= G_1(t) + \sum_{n=2}^{\infty} \int_0^t F(t-s)\,dG_{n-1}(s) \qquad \begin{array}{l}\text{using recurrence}\\ \text{relation for } G_n\end{array}$$

$$= G_1(t) + \int_0^t F(t-s)\,d\sum_{n=2}^{\infty} G_{n-1}(s) \qquad \text{change of order}$$

$$= G_1(t) + \int_0^t F(t-s)\,dU(s) \qquad \text{definition of } U(s)$$

Recall that $G_1 = F$, so we have the <u>renewal equation</u> for $U(t)$:

$$\boxed{U(t) = F(t) + \int_0^t F(t-s)\,dU(s)}.$$

This is the famous renewal equation -- it allows us to find $U(t)$ using only the known d.f. F -- thus avoiding finding individual G_n.

Define the <u>renewal density</u> $u(t) = dU(t)/dt$. Remember that $U(t)$ is the expected number of renewals up to time t (and it is <u>not</u> the d.f.), so $u(t)$ may be regarded as the expected number of renewals per unit of time; more precisely, the average number of renewals during the interval from t to t + h is

$$U(t+h) - U(t) = \int_t^{t+h} u(s) \ ds.$$

Differentiating the renewal equation one obtains the renewal equation for the density u:

$$u(t) = f(t) + \int_0^t f(t-s)u(s) \ ds, \qquad t > 0 \ .$$

Thus, knowledge of the density f of life times is sufficient to determine density u of renewals. This is a rather remarkable result. It has many applications in various fields.

Since the equation for u involves u also under the integral sign, the renewal equation has the form of an integral equation. Special techniques are needed for finding a solution. However, in the present case a solution can be found by a very simple method, namely the Laplace transform defined in Section 7.

Laplace transforms of densities f and u are defined by the following integrals:

$$f^*(\alpha) = \int_0^\infty e^{-\alpha t} f(t) \ dt, \qquad u^*(\alpha) = \int_0^\infty e^{-\alpha t} u(t) \ dt, \qquad \alpha \geq 0.$$

Write

$$w(t) = \int_0^t f(t-s)u(s) \ ds.$$

Simple evaluation of the double integral shows that the
Laplace transform of w(t) is

$$w^*(\alpha) \;=\; \int_0^\infty e^{-\alpha t} w(t) \; dt \;=\; f^*(\alpha) u^*(\alpha)$$

(This is expressed by saying that the Laplace transform of a
convolution is a product of Laplace transforms.)

Now taking Laplace transforms of both sides of the
renewal equation for density, one finds that

$$u^*(\alpha) \;=\; f^*(\alpha) + f^*(\alpha) u^*(\alpha).$$

Hence

$$\boxed{\; u^*(\alpha) \;=\; \frac{f^*(\alpha)}{1 - f^*(\alpha)} \;}.$$

Thus, to find the renewal density, we first compute the
Laplace transform of the density f, then calculate the
Laplace transform u^* from the above formula. It then
remains to invert $u^*(\alpha)$ to find u(t). This last step can
be facilitated by the use of tables of Laplace transforms.

Thus, we may consider our problem of finding the renewal
function U(t) as solved.

It may be proved that as time increases the renewal
density u tends to a constant value, equal to the recipro-
cal of the average life time $\mu = E(X)$. This is the famous
renewal theorem:

$$\boxed{\lim_{t\to\infty} u(t) \;=\; 1/\mu}\;.$$

Thus, whatever the life time distribution, the average number
of renewals up to t is approximately (for large t):

$$\boxed{U(t) \;\approx\; t/\mu}$$

and clearly tends to infinity as $t \to \infty$. This agrees with
the intuitive interpretation.

In another formulation, the average number of renewals
in the interval (t,t+h) for $h > 0$, is approximately for
large t:

$$\boxed{U(t+h) - U(t) \;\approx\; h/\mu}$$

for any form of life time distribution with mean life time
μ.

Remark: S_n is the total time for n renewals. Replacing
fixed n by a random variable N_t, one has a new random
variable S_{N_t} which represents the total time needed for
the random number of renewals.

Thus, $S_{N_t+1} - t$ is the time interval from an arbitrary
instant t (following a renewal) until the next renewal.
That is precisely the remaining life time since t. This is
the situation discussed in connection with a bus problem;
the distribution of the waiting time for the next renewal we

found in Section 4 corresponds to the distribution of $S_{N_t+1} - t$
as t goes to infinity (so called equilibrium conditions).

****

<u>Poisson Example</u>: The case of exponential life time and the
associated Poisson distribution (discussed in the previous
section) is of great importance in applications.

We have here

$$f^*(\alpha) = \int_0^\infty e^{-\alpha t} \lambda e^{-\lambda t} \, dt.$$

Hence

$$u^*(\alpha) = \lambda/\alpha$$

and this implies that $u(t) = \lambda$ is a constant. Hence the
limit is of course λ, too, and one has exactly:

$$u(t) = \lambda t.$$

In other words, $E(N_t) = \lambda t$, as it could be seen from the
fact that N_t has Poisson distribution.

To avoid confusion, let us repeat that the mean number
of renewals is λt, whereas $1/\lambda = \mu$ is the mean life time
(between renewals).

<u>Example 2</u>: Suppose that customers arriving to a shop (or
telephone calls at an exchange, insurance claims, etc.) can
be described by a renewal process. That is, let X_n be the

time interval between the arrival of the (n-1)-th and n-th customer. Suppose that the average interarrival time is μ = 5 min.

Then, approximately the average number of arrivals within one hour is t/μ = 60/5 = 12 (irrespective of the form of distribution of interarrival times).

Assume now that interarrival times have exponential density. Then, the pr. of having exactly 4 customers within a quarter of an hour is

$$P_{15}(4) = 3^4 e^{-3}/4! = 0.168$$

because λt = 15/5 = 3.

Example 3: Suppose that the average number of cars passing during a (12-hour) day is 360. Find the average number of cars during 2 hours. Here we have approximately

$$U(t+2) - U(t) \approx 2/\mu = 2(320/12) = 60$$

because μ = 12/360. What is the pr. of having exactly 60 cars in a 2 hour period, assuming a Poisson distribution? This can be read off from the tables, but in this range our tables are insufficient. One can use the Gaussian approximation, as in Section 9:

$$pr(59 < N_t \le 60) \quad = \quad pr(-1/8 < Z \le 0) \quad = \quad \Phi(1/8) - 1/2$$

$$= \quad 0.540 - 0.500 \quad = \quad 0.040,$$

recalling that the standard deviation is $\sqrt{60} \approx 8$.

Example 4: Suppose that the life time has uniform distribution with density $1/L$ over the interval $(0,L)$. Then

$$f^*(\alpha) \quad = \quad (1-e^{-\alpha L})/\alpha L$$

and

$$u^*(\alpha) \quad = \quad (1-e^{-\alpha L})/(e^{-\alpha L}-1+\alpha L).$$

It can be verified that

$$\lim_{\alpha \to 0} \alpha u^*(\alpha) \quad = \quad \frac{2}{L}$$

as it should. Hence, for $h > 0$

$$U(t+h) - U(t) \quad \approx \quad 2h/L.$$

Counterexample. As noted, given f determines u through the renewal equation. The converse need not be true, however. Indeed, the selection of arbitrary renewal density u, even if $u(t) \to 1/\mu$ as $t \to \infty$, does not imply in general that the inverted renewal equation

$$f^*(\alpha) \quad = \quad \frac{u^*(\alpha)}{1 + u^*(\alpha)}$$

should give a density. Indeed, suppose that

$$u(t) \;=\; (1-e^{-\mu t})/\mu \;\to\; 1/\mu \qquad\qquad \text{as} \quad t \to \infty.$$

The Laplace transform is

$$u^*(\alpha) \;=\; \frac{1}{\mu}\left(\frac{1}{\alpha} - \frac{1}{\alpha+\mu}\right)$$

and therefore

$$f^*(\alpha) \;=\; 1/(\alpha^2 + \mu\alpha + 1).$$

Suppose now that $\mu < 2$. Then, it can be verified by integration that the function

$$f(t) \;=\; e^{-\mu t/2}\,\frac{\sin(\zeta t/2)}{(\zeta/2)}, \qquad 0 < t < \infty,$$

where $\zeta = \sqrt{4-\mu^2}$, has the above $f^*(\alpha)$ as its Laplace transform. But of course, this $f(t)$ is not a density

(why?), although $\int_0^\infty f(t)\,dt = 1$.

Example 6: Let g be the density of the waiting time in the bus problem (Section 4) associated with the density f of the interarrival time. Let us take this waiting time as our life time X of the renewal process. It is easy to see that the Laplace transform of g is

$$g^*(\alpha) \;=\; \frac{1 - f^*(\alpha)}{\alpha\mu}$$

where μ is the mean of f. Then,

$$u^*(\alpha) = \frac{1 - f^*(\alpha)}{\alpha\mu - 1 + f^*(\alpha)} .$$

The renewal theorem yields:

$$\lim_{\alpha \to 0} \alpha u^*(\alpha) = \lim_{t \to \infty} u(t) = 1/w$$

where w is the average waiting time found in Section 4.
As a matter of fact we can use this observation to derive
the expression for w.

Section 18: Jumping Rabbit

A rabbit runs across field and suddenly falls into a
deep ditch. To get out, the rabbit jumps several times
until the magnitude of a jump is larger than the height of
the ditch. What is the pr. that the rabbit will be free on
the n-th attempt? What is the average number of attempts?
We tacitly assume that the rabbit is determined and does not
tire easily.

Many situations can be described by the jumping rabbit
model. Suppose that there is a certain critical level (like
a noise threshold, the level of water in a river, the boiling
temperature, the height of a ditch). Observations of a life
time are made and and long as the observed values stay below
the critical level, no action is taken. At the instant
when the critical level is reached, (for the first time),

an alarm is sounded. Of interest is the number of attempts
for the first occurrence of the alarm.

18.1.

Denote by X_n the magnitude of the n-th attempt (say,
the n-th jump), $n = 1,2,\ldots$, and let x be the critical
level (height of a ditch). As long as $X_n \leq x$ attempts
will continue, until the first overshot over x ($X_n = x$
means the rabbit's nose reached the surface; to get out the
rabbit needs more than x .

Let N be a r.v. representing the successful attempt of
the first jump over the level x. Clearly:

$$N = \min(n: n \geq 1, X_n > x).$$

Write for the distribution of N:

$$\mathrm{pr}(N = n) = P_x(n), \qquad n = 1,2,\ldots ,$$

where x is a parameter denoting the critical level.

To make the problem precise, further assumptions must
be imposed. The following ones are the simplest (although
not always too realistic):

A1: attempts are made independently of each other; that
     is, r.v.'s X_n are independent.

A2: the magnitude of each attempt is governed by the same
     distribution with density f and mean μ; that is,
     all life times X_n have the same distribution.

The indicated procedure implies that the event "rabbit is out at the n-th attempt" is:

$$(N = n) = (X_1 \leq x, X_2 \leq x, \ldots, X_{n-1} \leq x, X_n > x) \quad \text{for} \quad n = 2, 3, \ldots$$

$$(N = 1) = (X_1 > x) \qquad\qquad\qquad\qquad \text{for} \quad n = 1.$$

Hence, by independence:

$$\begin{aligned}
\mathrm{pr}(N = n) &= \mathrm{pr}(X_1 \leq x, \ldots, X_{n-1} \leq x, X_n > x) \\
&= \mathrm{pr}(X_1 \leq x) \ldots \mathrm{pr}(X_{n-1} \leq x)\mathrm{pr}(X_n > x) \\
&= F(x) \ldots F(x) F^c(x) = F^{n-1}(x) F^c(x), \quad \text{for} \quad n = 1, 2, \ldots
\end{aligned}$$

In other words we get the geometric distribution:

$$\boxed{P_x(n) = pq^{n-1}, \qquad n = 1, 2, \ldots}$$

where

$$p = F^c(x), \qquad q = F(x).$$

From Section 13 we know that the average number of attempts will be:

$$E(N) = 1/F^c(x)$$

and the pr. of some more than n attempts is $\mathrm{pr}(N > n) = F^n(x)$.

<u>Example</u>: (i) For the exponential distribution of rabbit's jumps:

$$p = e^{-\lambda x}, \qquad q = 1 - e^{-\lambda x}$$

so

$$P_x(n) = (1-e^{-\lambda x})^{n-1}e^{-\lambda x}, \qquad E(N) = e^{\lambda x}, \qquad pr(N > n) = (1-e^{-\lambda x})^n$$

ii) For the uniform distribution of rabbit's jumps:

$$p = 1 - x/L, \qquad q = x/L$$

where L is the maximum height of a jump, so

$$P_x(n) = (1-x/L)(x/L)^{n-1}, \qquad E(N) = L/(L-x), \qquad pr(N > n) = (x/L)^n$$

where $0 < x < L$; clearly if $x \geq L$, then $p = 0$ and the rabbit will never get out.

18.2.

Consider now a different version of a jumping rabbit. A rabbit starts from an initial position, say after a nap, jumps here and there executing a "random walk" with jumps of various magnitude. What is the pr. that a rabbit will fall in a ditch?

To simplify the discussion, we shall assume that jumps occur along a straight line. Let X_n be the magnitude of

the n-th jump, $n = 1, 2, \ldots$. Positive values of X_n
mean that a jump to the right occurred, and negative values
of X_n indicate a jump to the left. Assume that (X_n)
are i.i.d. with common d.f. F having density f and mean
μ. Suppose that the initial position of the rabbit is at
the origin, so $S_0 = 0$, and let S_n denote the rabbit's
position after n jumps:

$$S_n = X_1 + \ldots + X_n, \quad n = 1, 2, \ldots .$$

Observe that although (X_n) are independent, (S_n) are
dependent r.v.'s.

Select a fixed positive t, and take the interval
(t, ∞) as the "taboo set" representing a ditch. If $S_n \le t$
the rabbit enjoys free jumps; as soon as $S_n > t$ for the
first time, the rabbit fell into a ditch and the process
stops. If N is a r.v. representing the number of jumps to
reach the ditch, then clearly:

$$p(N = n) = pr(S_1 \le t, \ldots, S_{n-1} \le t, S_n > t).$$

Explicit determination of this pr. is tough, but several
general conclusions may be drawn easily. For convenience,
consider the event $A_n = (S_n > t)$, and note that its complement is $A_n^c = (S_n \le t)$. Hence we can write

$$pr(N = n) = pr(A_1^c \ldots A_{n-1}^c A_n).$$

Consider now the disjoint representation of a union of sets:
$\cup_{i=1}^{n} A_i = \cup_{i=1}^{n} A_1^c \ldots A_{i-1}^c A_i$. Hence

$$pr(\cup_{i=1}^{n} A_i) = \sum_{i=1}^{n} pr(A_1^c \ldots A_{i-1}^c A_i) = pr(N \le n)$$

or

$$pr(N > n) = pr(\cap_{i=1}^{n} A_i^c) = pr(S_1 \le t, \ldots, S_n \le t) \equiv G_n(t).$$

Thus, $G_n(t)$ denotes the pr. that the rabbit did not over-shoot t in the first n jumps, or equivalently that fall-ing into a ditch will occur after more than n jumps.

It is clear now that

$$pr(N = n) = G_{n-1}(t) - G_n(t), \quad \text{for} \quad n = 1, 2, \ldots$$

with $G_0(t) = 1$. Alternatively, this relation follows from the observation that:

$$pr(A_1^c \ldots A_{n-1}^c A_n) = pr(A_1^c \ldots A_{n-1}^c) - pr(A_1^c \ldots A_n^c).$$

Hence, by the same argument as in Section 17, the aver-age number of jumps for the rabbit is:

$$E(N) = \sum_{n=1}^{\infty} n\, pr(N = n) = \sum_{n=0}^{\infty} G_n(t) = 1 + \sum_{n=1}^{\infty} G_n(t).$$

And evidently, S_N is the position at which the poor rabbit landed in the ditch.

Recalling the definition of $M_n = \max(S_1, \ldots, S_n)$ from Section 6, one can see that

$$G_n(t) = \text{pr}(M_n \le t).$$

It should be stressed that G_n is the d.f. of M_n, and not of S_n (that was the case of renewals discussed in Section 16 because of positivity of life times). Indeed, the pr. of not crossing the barrier at t is the same as the pr. that the maximum of positions did not cross t.

However, it can be deduced that (see Problem 17)

$$G_n(t) = \int_{-\infty}^{t} G_{n-1}(t-s)\, dF(s), \quad n = 1, 2, \ldots$$

with $G_0 = 1$. In principle, the above recurrence relations could be used for evaluation, but calculations are prohibitive, and special methods are required.

Letting $n \to \infty$, one obtains:

$$G_\infty(t) = \text{pr}(N = \infty) = \text{pr}(S_n \le t \text{ for all } n)$$

This pr. is zero for most situations. If it is not zero, then it will satisfy the integral equation

$$G_\infty(t) = \int_{-\infty}^{t} G_\infty(t-s)\, dF(s).$$

The task of finding a solution of this integral equation is fascinating, but regretfully we cannot discuss it here (the transform method used earlier does not work here!). We shall limit ourselves to several comments which are rather intuitive, although their mathematical justification is very delicate.

It can be shown that if $\mu > 0$, the random walk of our rabbit will drift to $+\infty$ (visiting the positive half-axis infinitely often) and if $\mu < 0$, then the random walk will drift to $-\infty$ (visiting the negative half-axis infinitely often). When $\mu = 0$, then the random walk will oscillate, visiting positive and negative half-axes infinitely often.

Thus, for $\mu > 0$ the rabbit is bound to fall into the ditch (jump over t), so N will be finite, and this means that $G_\infty(t) = 0$ identically; in addition $E(N) < \infty$. As a matter of fact, one even has $E(S_N) = \mu E(N)$ as in Section 15.

On the other hand, for $\mu < 0$ the rabbit may fall into the ditch (having made some jumps to the right), or may not. Indeed, our rabbit may be lucky and never even approach

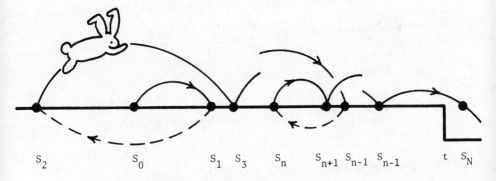

Fig 18.1
Jumping rabbit.

the ditch (stays below t). This means that N will assume
infinite value with positive probability. Thus, $G_\infty(t) =$
$pr(N = \infty) > 0$; hence also $E(N) = \infty$. As a matter of fact,
one has $G_\infty(\infty) = 1$, so G_∞ is the proper d.f.

Finally when $\mu = 0$, the hesitant rabbit will wander
along the field, and will eventually fall in. Thus, $G_\infty(t)$
$= 0$ again, so N is finite but then $E(N) = \infty$. It will
take awfully long on the average to reach the ditch if one
really does not care where one is going.

Chapter 3: Problems

1. Suppose that interarrival times between consecutive buses
    (arriving at a bus stop) form a renewal process; that is
    let X_n be a time interval between the (n-1)-th and
    n-th bus. Assume that the X_n are exponential life
    times with mean $\mu = 15$ minutes.

    Find the pr. of having two or more buses within
    a time interval of length t. Call this pr. b(t), and
    plot its graph as a function of t.

2. Mary has an electric gadget which works on a single
    battery. As soon as the battery in use fails, Mary
    immediately replaces it with a new battery.

    If the life time of a battery is (in hours) uni-
    formly distributed over the interval (20,40), then at
    what rate does Mary have to change batteries?

How many batteries on the average will Mary use per month (assume here that Mary uses the gadget only 5 hours per day)?

3. Peter buys records at the rate of 2 per month.

   (i) How many records will he buy during a year?

   (ii) Assuming the Poisson distribution, what is the pr. that during the summer (i.e., 2 1/2 months) he will buy

   (a) exactly 3 records,

   (b) more than 3 records,

   (c) at least one record?

4. Let $P_t(n)$ be the Poisson distribution with mean λt.

   (i) Show that:

   $$\int_0^\infty P_t(n) e^{-qt}\, dt = \lambda^n / (\lambda+q)^{n+1}$$

   where $q \geq 0$. For $q = 0$, the integral gives $1/\lambda$ for all n.

   (ii) Evaluate:

   $$\int_0^\infty t^m P_t(n)\, dt = \frac{(n+m)!}{n!} \frac{1}{\lambda^{m+1}}.$$

   (iii) Plot the graph of the function $P_n(t)$ for $t > 0$, for fixed $n = 0,1,\ldots$. Indicate the shape, find the maximum, inflection points, etc. (Consider separately $n = 0$ and $n \neq 0$).

5. Let $P_t(n)$ be the Poisson distribution with mean λt.
    Write $Q_n(t) = dP_t(n)/dt$. Show that

    (i) $Q_n(t) = \lambda P_t(n-1) - \lambda P_t(n)$ for $n \geq 1$, and
          $\qquad\qquad = -\lambda P_t(0)$ for $n = 0$.

    (ii) $\qquad\qquad \int_0^\infty Q_n(t)\ dt = 0$ for $n \geq 1$,
          $\qquad$ and $\qquad\qquad\qquad = -1$ for $n = 0$.

6. Show that for each k:

$$\lim_{t \to \infty} \sum_{n=k+1}^\infty \frac{(\lambda t)^n}{n!}\, e^{-\lambda t} = 1.$$

7. Suppose that the r.v. N_t has the Poisson distribution
    with mean λt. Let the cost function be $(-1)^n$, $n = 0,1,2,\ldots$. Show that the average cost is

$$\sum_{n=0}^\infty (-1)^n \frac{(\lambda t)^n}{n!}\, e^{-\lambda t} = e^{-2\lambda t}.$$

    Note: No calculation needed! Compare this with
    Example 3 in Section 14).

8. Let f be the density of a life time of the following
    form:

$$f(x) = \lambda \frac{(\lambda x)^{n-1}}{(n-1)!}\, e^{-\lambda x},\quad n = 1,2,\ldots,\quad x \geq 0$$

    Show that

$$\int_0^t f(x) e^{-\lambda(t-x)}\ dx$$

    gives the Poisson distribution with mean λt.

9. Suppose that the life time in a renewal process has uniform distribution over the time interval $(a, a+L)$ where a and L are positiive constants.

    Show that the average number of renewals during the interval $(t, t+h)$, with $h > 0$, is given approximately (for large t) by $2h/(2a+L)$.

10. Suppose that the life time in a renewal process has the d.f. of the form:

    $$F(t) = \frac{t^2}{(1+t^2)} \qquad \text{for} \quad t \geq 0.$$

    Show that the average number of renewals during the interval $(t, t+h)$, with $h > 0$, is given approximately (for large t) by: $2h/\pi$.

11. Suppose that the life time in a renewal process has d.f. of the following form:

    $$F(x) = 1 - e^{-\lambda\sqrt{x}} \qquad \text{for} \quad x \geq 0$$

    where $\lambda > 0$ is a constant.

    Show that the average number of renewals during the time interval $(t, t+h)$, with $h > 0$, is given approximately (for large t) by $\lambda^2 h/2$.

12. Suppose that the life time in a renewal process has density of the form

    $$f(t) = cae^{-at} + (1-c)be^{-bt}, \qquad \text{for} \quad t \geq 0$$

    where $a \neq b$, $a > 0$, $b > 0$ and $0 < c < 1$.

Show that the average number of renewals during
the interval $(t,t+h)$, with $h > 0$, is given approximately (for large t) by:

$$abh/(a+cb-ca).$$

13. Suppose that the life time in a renewal process has
density of the form

$$f(t) \;=\; \begin{cases} c\lambda e^{-\lambda t} + (1-c)\lambda, & \text{for } 0 < t < 1/\lambda \\ c\lambda e^{-\lambda t}, & \text{for } 1/\lambda < t < \infty, \end{cases}$$

where $\lambda > 0$ and $0 < c < 1$.

Show that the average number of renewals during
the interval $(t,t+h)$, with $h > 0$, is given approximately (for large t) by:

$$2\lambda h/(1+c).$$

14. Suppose that the density f of the common life time X
in the renewal process is supported by the interval
$[s,s+L]$, as in Section 2. Then, clearly the total
life times S_n take values in the interval $[ns,ns+nL]$.
   Show that:

$$pr(N_t \geq n) \;=\; \begin{cases} 0 & \text{for } t < ns \\ 1 & \text{for } t \geq n(s+L). \end{cases}$$

15. Jack keeps a gold fish in a fish bowl of height L,
and checks the level of water each morning. If the
level exceeds x inches (where $0 < x < L$), he does

nothing. If the level drops below x, he fills the bowl to capacity with water.

Suppose that the distribution of the water level is uniform on the interval (0,L). Show that the probability that the bowl is filled on the n-th day for the first time is:

$$p(n) = (x/L)(1 - x/L)^{n-1}, \quad n = 1,2,\ldots$$

and that the average number of days until the first fill is L/x.

16. In the jumping rabbit problem, let $P_x(n)$ be the pr. that the rabbit escapes at the n-th attempt, assuming exponential distribution of the rabbit jumps.

Show that for fixed $n \geq 2$, this pr. has the maximum value for the ditch depth x equal to:

$$x = \lambda^{-1} \log n.$$

What for n = 1 ?

17. Proceeding as in Section 5, show that $G_2(t)$ from Section 18.2 can be expressed as:

$$G_2(t) = pr(S_1 \leq t, S_2 \leq t) = \iint_{\substack{x \leq t \\ x+y \leq t}} f(x)f(y)\, dx\, dy$$

$$= \int_{-\infty}^{t} F(t-s)f(s)\, ds.$$

Then, proceeding by induction, derive the recurrence relation for $G_n(t)$ stated in Section 18.2.

18. Take $f(x) = \frac{1}{2}e^{-|x|}$ and using Problem 17 show that

$$G_2(t) = \begin{cases} (3/8)e^t & \text{for } t \leq 0 \\ 1 - (5/8)e^{-t} - (1/4)te^{-t} & \text{for } t \geq 0 \end{cases}$$

and $E(M_2) = 1/2$.

Try for $G_3(t)$, if you have the patience.
Note, however, that now $\mu = 0$, so necessarily
$G_\infty(t) = 0$ for all t.

19. Suppose that jumps X_n in Section 18.2 have density of
the form:

$$f(t) = \begin{cases} \dfrac{\lambda\mu}{\lambda+\mu}\,e^{\lambda t}, & \text{for } t < 0 \\[2mm] \dfrac{\lambda\mu}{\lambda+\mu}\,e^{-\mu t}, & \text{for } t > 0 \end{cases}.$$

where $\lambda > 0$ and $\mu > 0$ are constant and $\rho = \lambda/\mu < 1$.
Hence $E(X_n) = (\rho-1)/\lambda < 0$ (actually, X_n is the
difference of two exponential life times).

Verify that the d.f. G_∞ defined by:

$$G_\infty(t) = \begin{cases} (1-\rho)e^{\lambda t}, & \text{for } t \leq 0 \\[2mm] 1 - \rho e^{-\mu(1-\rho)t}, & \text{for } t \geq 0 \end{cases}$$

satisfies the integral equation

$$G_\infty(t) = \int_{-\infty}^{t} G_\infty(t-s)f(s)\,ds$$

for all t.

Verify that G_∞ has a density which is a continuous function of t.

Chapter 4

Markovian Dance

In many practical situations we are interested in random fluctuations of a number of elements or components of a certain kind. Typically, we may investigate the number of people in the waiting line, or the number of customers in a shop, or the number of animals in a certain region. The discrete life time which expresses this number depends itself on time, so its probability distribution will also be a function of time. In other words, we have a family of discrete r.v.'s, say Y_t, where t is time, and this family $(Y_t, 0 \leq t < \infty)$ is called a stochastic process (with discrete state space and continuous parameter).

There is a class of stochastic processes which is characterized by some kind of lack of memory, in the sense that the future development of the process depends only on the present, but not the past. There are many important practical problems which have (or may be thought to have) this property. You may recall that our old friend the exponential life time has it. Such processes are known as Markov processes, and we shall study in this Chapter many interesting problems in which fluctuations dance to a Markovian tune.

Actually, we shall consider examples which belong to a nar-

rower category, appropriately called the "birth and death"

process.   Indeed, we regard increase in the number as births

and decrease as deaths (think about a colony of bacteria).

These processes have been investigated extensively in recent

years both from a theoretical point of view and for practi-

cal applications.

In Section 19 we shall examine the Poissonian input and

get another derivation of the Poisson distribution.   Section

20 treats the learning model and Section 21 a group of lines;

both sections hint at another point of view (related to

renewals).   Sections 22 and 23 treat the simple queueing

model of great importance in applications.   Section 24 deals

with general births, exclusively.   The method of analysis is

that of difference-differential equations for transition

pr.'s.   These equations are derived individually for each

case.   This seems to be very instructive.   However, in Sec-

tion 25 the general theory of birth and death processes is

illustrated from the modern point of view, based on the

"balance equation" (Chapman-Kolmogorov), and the reader is

probably left baffled by all that stuff.   We could probably

begin this chapter with Section 25 and derive others from it.

But in this way we may lose a lot of charm peculiar to the

individual special cases.   So it is better to leave Section

25 at the end, and content ourselves with a brief indication

of what is in store by alluding to strange things in the

last-by-one Section 24. The reader who wishes to investigate
matters further is strongly encouraged to look at Section 25.

Again, this chapter leans heavily on the previous
chapters, but its mathematics is perhaps a little more difficult. And, unfortunately, calculations become lengthy, too.
But that is the nature of our subject, and we may as well
start to like it.

Section 19: Poisson Input

Consider a "system" (a shop, a news stand, an office,
a factory, a telephone exchange, etc.) to which "customers"
arrive (people, machines, letters, messages, claims, calls,
etc.). We shall be interested in the number of customers
who arrived within a given time interval -- i.e., the input
to the system. Actually, we shall consider a most popular
kind of input, known as the Poisson input.

Denote by Y_t a number of customers present in the
system at time t. Here $0 \leq t$, and Y_0 is the number of
customers present initially. Clearly, $Y_t - Y_0 = N_t$ is the
number of those who arrived during the time interval $(0,t)$.
Similarly, the number of arrivals during time interval
$(t,t+h)$ is $Y_{t+h} - Y_t$. Poisson input is specified by the
following assumptions:

A1: The pr. that during an interval $(t,t+h)$ where $h > 0$
     is small, an arrival occurs is approximately equal to
     λh.

A2:   The pr. of more than one arrival during (t,t+h) is
      negligible.

Physically, this implies that the total number of incoming
customers can only increase in unit steps, with the pr. of
a new customer independent of time and of the number of
customers present, and proportional to the interval length
h. The factor of proportionality λ is called the intensity,
or <u>arrival rate</u>.

      We shall be interested in the conditional pr. that at
time t there are j customers in the system, when at time
0 there have been i customers. In symbols:

$$p_{ij}(t) \;=\; pr(Y_t = j \mid Y_0 = i), \qquad 0 \le t, \quad i,j = 0,1,2,\ldots .$$

The event $(Y_t = j)$ is frequently expressed by saying that
at t the system is in state j. Thus, $p_{ij}(t)$ expresses
the <u>transition pr.</u> from state i (initially) to state j
at time t. Since the number of customers can only increase,
it is clear that

$$p_{ij}(t) = 0 \qquad \text{for } i > j.$$

      Assumptions Al - 2 are used to derive equations for the
transition pr. For this purpose, consider two contiguous
intervals (0,t) and (t,t+h), where h is small. Suppose
that at t = 0 there are initially i customers, and at
time t+h there are j. Clearly j ≥ i. If at least one
arrival occurred during the time interval (0,t+h), then

one must distinguish the three mutually exclusive ways for this to happen:

(a)   no arrival during (t,t+h) and j-i arrivals during (0,t), i.e., transition from i to j;

(b)   one arrival during (t,t+h) and j-i-1 arrivals during (0,t), i.e., transition from i to j-1;

(c)   two or more arrivals during (t,t+h) and less than j-i-1 arrivals during (0,t), i.e., transition from i to less than j-1.

In terms of pr.'s the above contingencies may be expressed by the "balance equation":

$$p_{ij}(t+h) \;=\; p_{ij}(t)p_{jj}(h) + p_{ij-1}(t)p_{j-1,j}(h) + \cdots$$

The pr. of the third contingency being negligible by A2. Now, by A1 one has approximately

$$p_{j-1,j}(h) \;=\; \lambda h.$$

Hence, the pr. of no arrival during (t,t+h) is: $p_{jj}(h) = 1 - \lambda h$, approximately. Consequently, the balance equation becomes (up to approximation):

$$p_{ij}(t+h) = p_{ij}(t)(1-\lambda h) + p_{ij-1}(t)\lambda h + \cdots, \qquad \text{for small} \quad h.$$

Hence,

$$\frac{p_{ij}(t+h) - p_{ij}(t)}{h} \;=\; -\lambda p_{ij}(t) + \lambda p_{ij-1}(t) + \cdots .$$

Passing to the limit as $h \to 0$, one obtains a difference-differential equation:

$$\frac{dp_{ij}(t)}{dt} = -\lambda p_{ij}(t) + \lambda p_{ij-1}(t), \qquad \text{for} \quad j > i \qquad (*)$$

For $j = i$, only the first contingency can occur, so by the similar reasoning

$$p_{ii}(t+h) = p_{ii}(t)(1-\lambda h) + \dots$$

which leads to:

$$\frac{dp_{ii}(t)}{dt} = -\lambda p_{ii}(t), \qquad \text{for} \quad j = i \qquad (**)$$

The initial condition is expressed by:

$$p_{ii}(0) = 1, \qquad p_{ij}(0) = 0 \quad \text{for} \quad j \neq i \qquad (***)$$

There are several methods for solving this infinite system of difference-differential equations. The most natural one, although not the simplest, is the step-by-step method. Starting with the differential equation (**) and using the initial condition (***), it is easily seen that

$$p_{ii}(t) = e^{-\lambda t}, \qquad \text{for} \quad t \geq 0.$$

Remark: Observe that the above expression is in fact the pr. of no arrival up to t, that is of no change in state

i. In other words, this is the complementary d.f. of the exponential life time (the interarrival period). This is in agreement with the discussion of the Poisson process in terms of renewals. Furthermore, one has for small h:

$$e^{-\lambda h} = 1 - \lambda h + \dots$$

indicating that A1 indeed corresponds to the pr. of termination being approximately λh. See Section 3.

The next step is the solution for $j = i + 1$. One has here the linear differential equation

$$\frac{dp_{i,i+1}(t)}{dt} + \lambda p_{i,i+1}(t) = \lambda e^{-\lambda t}, \quad \text{with } p_{i,i+1}(0) = 0.$$

Then, for $j = i + 2$, and so on. It is, however, much simpler to observe that differentiation of the Poisson distribution with respect to t, carried out in Section 17, yields equations analogous to (*) and (**). Thus, one can easily verify by differentiation that the solution is:

$$p_{ij}(t) = \frac{(\lambda t)^{j-i}}{(j-i)!} e^{-\lambda t}, \quad \text{for } j \geq i, \quad t \geq 0.$$

Thus, we have again obtained the Poisson distribution; hence the name-Poisson input. Note that this is actuallly the third derivation of the Poisson distribution; in fact this is the most common one.

****

Suppose that we are given the initial distribution of Y_0: $\mathrm{pr}(Y_0 = i) = \pi_i$. Then the joint distribution of Y_0 and Y_t is (by rules of conditional pr.):

$$\mathrm{pr}(Y_0 = i, \ Y_t = j) \ = \ \pi_i p_{ij}(t).$$

Hence, the distribution of $N_t = Y_t - Y_0$, the number of arrivals during $(0,t)$ is:

$$\mathrm{pr}(N_t = n) \ = \ \sum_{i=0}^{\infty} \pi_i p_{i,i+n}(t) \ = \ P_t(n)$$

where $P_t(n)$ is the Poisson distribution as in Section 16, in agreement with the definition of N_t.

$$*********************$$

The <u>conditional</u> <u>expectation</u> of Y_t given that $Y_0 = i$, is defined by

$$E(Y_t \mid Y_0 = i) \ = \ \sum_{j=0}^{\infty} j p_{ij}(t).$$

This is the average number of customers present at time t, given that initially there were i customers. Easy computation shows that the above conditional expectation is:

$$\sum_{j=i}^{\infty} [(j-i)+i] p_{ij}(t) \ = \ \sum_{n=0}^{\infty} n P_t(n) + i \sum_{j=i}^{\infty} p_{ij}(t) \ = \ \lambda t + i$$

which is intuitively obvious, as $E(N_t) = \lambda t$ is the expected number of arrivals during $(0,t)$ and i is the initial num-

ber. Note that the conditional expectation is not a number,
but a function of i.

Example: If the number of accidents since the beginning of
this week has been 4, and the rate is 7 accidents per week,
what is the pr. of an accident tomorrow? Here $i = 4$, $j =$
5 and $\lambda = 7/7 = 1$ per day, and $t = 1$ day. So $\lambda t = 1$,
$n = j-i = 1$, and $P_1(1) = 0.3679$.

   The average number of accidents since the beginning of
the week until tomorrow will be $i + \lambda t = 4 + 1 = 5$.

****

Random time:

   Consider again the Poisson input, and assume for convenience that initially (at $t = 0$) there were no arrivals
($i = 0$). Thus, N_t is the number of arrivals during $(0,t)$,
with $N_0 = 0$, and the distribution of N_t is Poissonian
with mean λt, namely $p_{0j}(t)$ for $j = 0,1,\ldots$.

   Consider now, instead of an interval of fixed (determin-
istic) length t, an interval of random length T. Here T
is a life time of an interval. We would like to find the
distribution of arrivals during such random intervals. We
shall write N_T in analogy to N_t, to stress the fact that
duration of the interval is now T. We shall assume that T
is independent of the Poisson input.

   This situation is very common in applications. For
example, in the telephone system T may represent the "hold-

ing time" of a call (or duration of a conversation), and it is required to find the distribution of incoming calls during a holding time. In other application, T may be duration of the observation interval (for road traffic, say) and N_T gives the number of arrivals during that period.

As N_t is a well behaved (discrete) life time, and T a life time well familiar to us, we may rest assured that N_T will also be an honest (discrete) life time. Indeed, when T = t, then N_T becomes N_t. Hence, we can regard distribution of N_t as that of N_T, but conditional on T = t:

$$pr(N_T = j \mid T = t) = \frac{(\lambda t)^j}{j!} e^{-\lambda t} = pr(N_t = j).$$

Let f be the density of the life time T. Then, by a conditioning argument (similar to that in Section 15), the distribution of N_T is given by:

$$pr(N_T = j) = \int_0^\infty pr(N_T = j \mid T = t) \, f(t) \, dt$$

$$= \int_0^\infty \frac{(\lambda t)^j}{j!} e^{-\lambda t} \, f(t) \, dt.$$

This is the required formula.

It is easy to see that the average number of arrivals is now:

$$E(N_T) = \lambda E(T)$$

in agreement with intuition; indeed, t in λt is now replaced by $E(T)$ -- the average (see also a similar relation in Section 15).

As an example, suppose that f is exponential with parameter μ. Then

$$\mathrm{pr}(N_T = j) = \int_0^\infty \frac{(\lambda t)^j}{j!} e^{-\lambda t} \mu e^{-\mu t} \, dt = \frac{\mu \lambda^j}{(\mu + \lambda)^{j+1}},$$

$$j = 0, 1, \ldots,$$

which is of course the geometric distribution with mean λ/μ (see also Section 22).

Section 20: It Is Easy to Learn

We shall now consider a simple learning model in which an individual learns to perform one action correctly. We shall assume that when the action is performed, the individual's response may be either correct or incorrect. As time progresses, one obtains a sequence of responses; if from some time on, the correct responses prevail, we may consider this as an indication that the individual learned to perform the assigned task. If incorrect responses prevail, the individual did not learn anything. We assume that time flows continuously, but tasks (i.e., actions) are performed successively at some aribitrary instants.

It will be convenient to denote responses by 0 and 1, and classify them as

0 = correct response,   1 = wrong response.

Denote by Y_t the r.v. which indicates the response which

prevails at time   t.   Thus, $(Y_t = 0)$ is the event that at
time   t   the correct response prevails; similarly $(Y_t = 1)$
is the event of incorrect response at time   t.   For simplification we shall say the system is in state   0   or in state 1
at   t.

We shall be interested in the pr. of state   j   at time
t,   given that initially state   i   prevailed:

$$p_{ij}(t) \;=\; pr(Y_t = j \mid Y_0 = i),$$

where   i   and   j   can be only   0   and   1; $t \geq 0$.   For
example, $p_{10}(t)$ is the pr. of a correct response at time
t,   when initially the response was wrong. To make the
problem more precise further assumptions are needed. We
shall again consider what may happen within a short interval
(t,t+h). We shall assume that during this interval, approximately:

pr. of correct response, when the wrong one prevailed at
t,   is $p_{10}(h) = ah + \ldots$

pr. of wrong response, when the correct one prevailed at
t,   is $p_{01}(h) = bh + \ldots$.

Coefficient   a   represents the learning rate; it would be
desirable to have   a   large. Coefficient   b   represents
the hindrance rate; it would be desirable to have   b   small.
One could say that a   is a measure of brightness, whereas
b   is a measure of dumbness.

To obtain equations for transition pr.'s $p_{ij}(t)$, we shall consider two contiguous intervals $(0,t)$ and $(t,t+h)$, and enumerate all possibilities of transitions. The following equations are self-evident -- they represent the "balance of pr.":

$$p_{00}(t+h) = p_{00}(t)p_{00}(h) + p_{01}(t)p_{10}(h)$$

$$= p_{00}(t)[1-bh] + p_{01}(t)ah + \ldots$$

$$p_{01}(t+h) = p_{00}(t)p_{01}(h) + p_{01}(t)p_{11}(h)$$

$$= p_{00}(t)bh + p_{01}(t)[1-ah] + \ldots$$

$$p_{10}(t+h) = p_{10}(t)p_{00}(h) + p_{11}(t)p_{10}(h)$$

$$= p_{10}(t)[1-bh] + p_{11}(t)ah + \ldots$$

$$p_{11}(t+h) = p_{11}(t)p_{11}(h) + p_{10}(t)p_{01}(h)$$

$$= p_{11}(t)[1-ah] + p_{10}(t)bh + \ldots$$

Hence, as in the previous section, putting $p_{ij}(t+h) - p_{ij}(t)$ on the left side, dividing by h, and passing to the limit $h \to 0$, one obtains the following system of differential equations:

$$\frac{d}{dt} p_{00}(t) = -bp_{00}(t) + ap_{01}(t) \qquad (1)$$

$$\frac{d}{dt} p_{01}(t) = bp_{00}(t) - ap_{01}(t) \qquad (2)$$

$$\frac{d}{dt} P_{10}(t) = - bP_{10}(t) - aP_{11}(t) \tag{3}$$

$$\frac{d}{dt} P_{11}(t) = bP_{10}(t) - aP_{11}(t) \tag{4}$$

with initial conditions:

$$P_{00}(0) = P_{11}(0) = 1, \qquad P_{01}(0) = P_{10}(0) = 0. \tag{5}$$

Observe also that necessarily,

$$P_{00}(t) + P_{01}(t) = 1, \qquad P_{10}(t) + P_{11}(t) = 1 \tag{6}$$

because possibilities of getting 0 or 1 are mutually exclusive. Note that the addition of equations (1)-(4) sidewise gives 0 = 0, as it should.

To solve the system, consider equation (1):

$$\frac{d}{dt} P_{00}(t) + bP_{00}(t) = aP_{01}(t) \tag{1}$$

and differentiate

$$\frac{d^2}{dt^2} P_{00}(t) + b\frac{d}{dt} P_{00}(t) = a\frac{d}{dt} P_{00}(t) = -a\frac{d}{dt} P_{00}(t)$$

using (2). So

$$\frac{d^2}{dt^2} P_{00}(t) + (a+b)\frac{d}{dt} P_{00}(t) = 0 \tag{1a}$$

Put $y(t) = \frac{d}{dt} P_{00}(t)$, so (1a) becomes

$$\frac{d}{dt} y(t) + (a+b) y(t) = 0 \qquad (1b)$$

and its solution is

$$y(t) = Ae^{-(a+b)t}. \qquad (1c)$$

Integrating (1c) again, one has

$$p_{00}(t) = B - \frac{A}{a+b} e^{-(a+b)t} \qquad (1d)$$

where   A   and   B   are constants. These constants are determined from initial conditions (5):

$$p_{00}(0) = 1, \qquad \frac{d}{dt} p_{00}(t) \bigg|_{t=0} = ap_{00}(0) - bp_{00}(0) = -b.$$

Hence, letting   t = 0   in (1c) and (1d):

$$A = -b, \qquad 1 = B + \frac{b}{a+b} \quad \text{or} \quad B = \frac{a}{a+b}$$

one has:

$$\boxed{p_{00}(t) = \frac{a}{a+b} + \frac{b}{a+b} e^{-(a+b)t}} \qquad (7)$$

Hence, from (6):

$$\boxed{p_{01}(t) = \frac{b}{a+b} - \frac{b}{a+b} e^{-(a+b)t}}. \qquad (8)$$

Similar analysis applied to equation (3) yields:

$$P_{10}(t) = \frac{a}{a+b} - \frac{a}{a+b} e^{-(a+b)t} \qquad (9)$$

$$P_{11}(t) = \frac{b}{a+b} + \frac{a}{a+b} e^{-(a+b)t} \qquad (10)$$

Suppose that we do not know what the initial values at $t = 0$ are. Then, the initial distribution must be given:

$$pr(Y_0 = 0) = \pi_0, \qquad pr(Y_0 = 1) = \pi_1, \qquad \text{with} \quad \pi_0 + \pi_1 = 1.$$

The joint distribution is then

$$pr(Y_0 = i, \ Y_t = j) = \pi_i P_{ij}(t), \qquad \text{with} \quad i,j = 0,1.$$

Hence the distribution at t is:

$$pr(Y_t = j) = \pi_0 P_{0j}(t) + \pi_1 P_{1j}(t), \qquad \text{for} \quad j = 0,1.$$

Denote this pr. by $\pi_j(t)$. Then, in our case

$$\pi_0(t) = \frac{a}{a+b} + \frac{\pi_0 b - a\pi_1}{a+b} e^{-(a+b)t} \qquad (11)$$

$$\pi_1(t) = \frac{b}{a+b} - \frac{\pi_0 b - a\pi_1}{a+b} e^{-(a+b)t} \qquad (12)$$

Note that $\pi_0(t) + \pi_1(t) = 1$, for all t, as it should. Also, (11) and (12) reduce to π_0 and π_1, respectively, for $t = 0$.

More important, however, is the fact that the limiting distributions coincide:

$$\lim_{t\to\infty} p_{i0}(t) \;=\; \lim_{t\to\infty} \pi_0(t) \;=\; \frac{a}{a+b} \;,$$

(13)

$$\lim_{t\to\infty} p_{i1}(t) \;=\; \lim_{t\to\infty} \pi_1(t) \;=\; \frac{b}{a+b} \;.$$

This means that as time progresses the system tends to an equilibrium -- that is states 0 and 1 are achieved with constant pr.'s, which are independent of the initial state. In other words, influence of the initial conditions wears out, and the system operates fluctuating between 0 and 1. For the learning model, this means that in the long run the pr. of correct responses is $\frac{a}{a+b}$; that of wrong responses is $\frac{b}{a+b}$.

Note that if a and b are chosen such that $\pi_0 b = \pi_1 a$, then $\pi_0(t)$ and $\pi_1(t)$ are constant for each t. Thus, for this special choice of initial distribution the system would behave as if it were already in equilibrium.

Indeed, the required condition can be satisfied iff the initial distribution (π_0, π_1) coincides with the limiting distribution (13).

It is of interest to plot graphs of $\pi_j(t)$. First of all note that:

$$\pi_0 \;<\; \frac{a}{a+b} \qquad \text{iff} \quad \pi_0 b - a\pi_1 < 0;$$

$$\pi_0 \;>\; \frac{a}{a+b} \qquad \text{iff} \quad \pi_0 b - a\pi_1 > 0.$$

Hence the graph of the pr. of correct response at time t is:

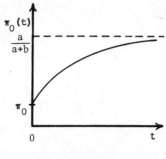

Fig 20.1

Probability of the correct response for π_0 < a/(a+b).

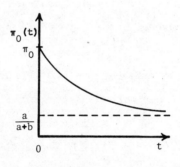

Fig 20.2

Probability of the correct response for π_0 > a/(a+b).

Section 21: On-Off Transitions

21.1.

Consider a single device (a bulb, machine, signal, neuron, etc.) which can only be in two states, say, on and off (busy-idle, good-bad, in-out, etc.). For convenience, we shall denote

state "off" as state 0, and state "on" as state 1.

The device may spend some time in each state, and as time progresses busy and idle periods alternate. Let

X_0 be a life time representing the idle period (i.e., when state 0 prevails),

X_1 be a life time representing the busy period (i.e., when state 1 prevails).

The life time $X = X_0 + X_1$ represents a cycle of repetition, that is the time interval between two consecutive instants of the beginning of state 1 (interarrival time) or between two consecutive instant of the beginning of state 0 (termination time). See the Figure below.

We shall now assume that X_0 and X_1 are independent, and denote their density, d.f. and mean by f_0, F_0, μ_0 and f_1, F_1, μ_1, respectively. Similarly, let f, F and μ correspond to the cycle life time X.

As shown in Section 5, the density f is given by convolution of densities f_0 and f_1:

$$f(t) = \int_0^t f_0(s) f_1(t-s) \, ds$$

and clearly $\mu = \mu_0 + \mu_1$.

One could regard a sequence of cycles X as a renewal process (so called alternating process). Let u be the corresponding density of renewals. We have now the renewal equation

$$u(t) \ = \ f(t) \ + \int_0^t u(t-s)f(s) \ ds$$

and by the renewal theorem (see Section 17) the limiting
rate is $1/\mu$. Thus the average number of arrivals, as
well as of terminations, in time interval $(0,t)$ is approximately t/μ.

From now on, we shall assume that life times X_0 and
X_1 are exponential with densities:

$$f_0(t) = be^{-bt}, \quad f_1(t) = ae^{-at}, \quad \text{for } t \geq 0, \ a > 0, \ b > 0.$$

Then:

$$f(t) \ = \ \int_0^t be^{-bs}ae^{-a(t-s)}ds \ = \ \frac{ab}{b-a} \, (e^{-at} - e^{-bt}) \ \geq \ 0$$

$$(t \geq 0)$$

for $a \neq b$; for $a = b$ see Example 1 in Section 5. The
d.f. is

$$F(t) \ = \ 1 - \frac{1}{b-a} \, (be^{-at} - ae^{-bt}), \qquad t \geq 0.$$

Clearly, $\mu = \frac{1}{a} + \frac{1}{b}$.

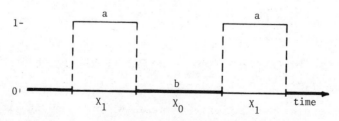

Fig 21.1

On-off transitions.

Now we shall consider another problem. Suppose that
we wish to find the pr. that at some instant t, a device
will be in state 0 or in state 1. Let Y_t be a r.v.
representing the state at t. Write again

$$p_{ij}(t) = pr(Y_t = j \mid Y_0 = i)$$

for the conditional pr. of state j at t, given that
initially at t = 0 state i prevailed. Here i and j
can assume only values 0 and 1.

A moments reflection shows that these transition pr.'s
$P_{ij}(t)$ are the same as those found in the previous Section
20. However, we shall verify this statement by a different
argument which further illustrates the connection between
the life times we discussed earlier, and the r.v.'s representing a number of elements (here 0 and 1, only).

Suppose that the device is initially in state 0 (that
is "off"). We wish to find the pr. of it being in state 1
(i.e. "on") at t. Thus, the device is at 0 initially,
and state 0 prevails until some instant s where the first
transition to state 1 takes place. This occurs with density
f_0 (termination of the idle state). Then during the remaining time interval t - s transitions from state 1 to state
1 may occur, with pr. $p_{11}(t-s)$. Hence, the total pr. is:

$$p_{01}(t) = \int_0^t f_0(s)p_{11}(t-s)\ ds.$$

If it is required that at t the device be in state 0, when initially it has been in 0, two possibilities must be considered. The device may stay in 0 up to time t, with no change; this means that life time X_0 continues beyond t, and the pr. of such an event is clearly $F_0^c(t) = e^{-bt}$. The second possibility is that a device originally at 0, changes to 1 and at some instant s returns for the first time to state 0; this corresponds to the cycle X with density f. Then, during the remaining time interval t - s some transitions from 0 to 0 may occur, with pr. $P_{00}(t-s)$. The total pr. is therefore:

$$P_{00}(t) = F_0^c(t) + \int_0^t f(s)p_{00}(t-s)\ ds.$$

It can be easily verified by substitution that expressions for the transition pr. $p_{ij}(t)$ found in the previous section satisfy the above equations.

By a similar argument, one obtains for the initial state 1:

$$P_{10}(t) = \int_0^t f_1(s)p_{00}(t-s)\ ds$$

$$P_{11}(t) = F_1^c(t) + \int_0^t f(s)p_{11}(t-s)\ ds.$$

****

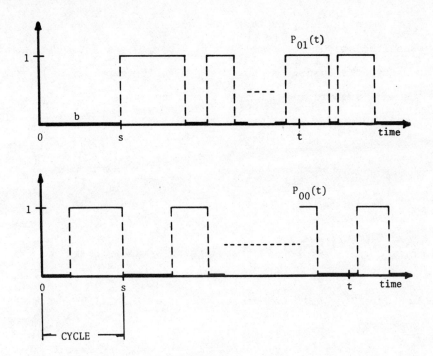

Fig 21.2

Transition probabilities.

21.2. Group of lines

Consider now a group of n lines, each of which may be busy or free, the change of state being governed by equations developed in this section (and in Section 20). Assume also that the individual lines act independently. We shall be now interested in the distribution of the number of busy lines in the group. Let Y_t be the number of busy lines at time t, with Y_0 being the initial number. We wish to find

$$p_{ij}(t) \ = \ pr(Y_t = j \mid Y_0 = i)$$

where now i and j range from 0 to n.

We previously treated the case $n = 1$. It is hoped
that the use of the same notation in this new set will not
confuse the issue. To avoid any misunderstanding, we shall
use new symbols for the two transition pr.'s which we shall
need, namely we shall write, respectively,

$$c_0(t) \quad \text{and} \quad c_1(t) \quad \text{for} \quad p_{01}(t) \quad \text{and} \quad p_{11}(t)$$

from subsection 21.1.

Thus, $c_0(t)$ is the pr. that a single line will be busy at
t, when initially it was free, and $c_1(t)$ is the pr. that
a single line will be busy at t, when initially it was
busy, too.

Initially, at $t = 0$ we have i busy lines and $n - i$
free lines, and at time t we wish to have j busy lines.
This change can be accomplished when originally free lines
become busy, and originally busy lines remain busy, at least
some of them. We must consider all possibilities.

Suppose, then that out of the $n - i$ free lines initially, k will become busy at t. We have here the Bernoulli
trials with pr. of "success" being $c_0(t)$. Hence, the
required pr. is:

$$q_i(k,t) \;=\; \binom{n-i}{k} [c_0(t)]^k [1-c_0(t)]^{n-i-k}$$

(see Section 11 and the somewhat analogous situation with
the rabbit in Section 18.1).

Furthermore, as there should be j busy at t so $j-k$ must remain busy from the initial i busy. Thus, again we have Bernoulli trials with pr. of "success" $c_1(t)$. The required pr. is

$$r_i(j-k,t) = \binom{i}{j-k}[c_1(t)]^{j-k}[1-c_1(t)]^{i-j+k}.$$

Furthermore, transitions from free to busy and from busy to busy on these lines, are independent, so in order to get transition from i to j, we should multiply the above pr.'s, and then add them through all possible intermediary k. Thus, our final result is:

$$p_{ij}(t) = \sum_k r_i(j-k,t)q_i(k,t), \qquad 0 \le t,$$

where actually the summation is taken over the range $\max(j-i,0) \le k \le \min(n-i,j)$.

It is easy to see that the (conditional) average number of busy lines is:

$$E(Y_t \mid Y_0 = i) = (n-i)c_0(t) + ic_1(t) = \frac{nb}{a+b} + (1 - \frac{nb}{a+b}) e^{-(a+b)t}$$

and tends to $\frac{nb}{a+b}$ when $t \to \infty$. (Evidently, when $n = 1$ the above results agree with those in Section 20 --conversely, the present situation of n > 1 corresponds to n independent learners.

Remark: Extension of analysis from the previous subsection makes the validity of the following equations connecting transition pr.'s with first passage distributions plausible:

$$p_{ij}(t) = \int_0^t f_{ij}(s) p_{jj}(t-s)\ ds \quad \text{for } i \neq j$$

$$p_{jj}(t) = e^{-q_j t} + \int_0^t f_j(s) p_{jj}(t-s)\ ds$$

where $q_j = (n-j)b + ja$, and f_{ij} is the density of time to first entrance to j (from i) and f_j is the density of the time of the first return to j.

****

Finally, it can be verified that the limiting distribution

$$\lim_{t \to \infty} p_{ij}(t) = P(j)$$

is independent of the initial state i, and is given by the binomial distribution with parameters n and b/(a+b).

Section 22: Queueing

Consider a "server" who performs a certain service for customers, like a clerk in a Post Office selling stamps, a doctor seeing patients, a mechanic servicing cars; a car washing facility, a bus stop with passengers waiting for a bus, and so on. Customers may arrive in any fashion what-soever. Then the server is free, the incoming customer is given service immediately. Customers arriving when the server is busy, are put in a line and wait. When a server completes the service, any customer from the waiting line may be given access to the server, depending on the rules of "queue discipline." The most typical is "first come first serve," although other possibilities have been considered too. For example, some priorities may be assigned to wait-ing customers, the "last come first served" being not uncommon. On the other hand, customers may just scramble for a server ("random servicing").

Of primary interest is the distribution of the custom-ers present in the system, and the distribution of the waiting time to obtain service. We shall consider this problem here under very specific assumptions, which reasonably well repre-sent the real life situation. The queueing system to be discussed here is known as a simple queue.

Structure of the system:

    (i)   only one server;

    (ii) when the server is busy, a waiting line is formed which may be infinite in length;

    (iii) customers are served on the "first come first served" basis (strict order servicing);

(iv)   no waiting line is possible when the server is
       free;

(v)   customers in the queue wait till serviced, then
       depart (no early departures prior to service).

Probabilistic properties of the simple queue are described
by the following assumptions:

(vi)   incoming customers form a Poisson input, with
       rate λ;

(vii) service times of customers are independent identi-
       cally distributed exponential life times, with
       common mean $1/\mu$;

(viii) service times and interarrival times are indepen-
       dent of each other.

Let Y_t be a r.v. representing a number of customers
present in the system at the instant t, including a
customer being served, if any. Clearly, Y_t assumes values
0 (meaning no customers), 1 (necessarily a customer being
served), 2 (one served, one waiting), and so on. Thus,
$(Y_t = j)$ is the event that at time t there are exactly j
customers in the system; $j-1$ of them are waiting, if any.
The event $(Y_t = j)$ is expressed by saying that the system
is in state j at t.

Customers arrive to the system, join the queue and
wait; then are served for the duration of their service time,
and depart afterwards. During their waiting and service,

other customers arrive. Thus, their total number Y_t
fluctuates as time t progresses.

We shall be interested in the conditional pr. that at
time t there are j customers in the system, when at time
0 there have been i customers (the initial state i at
$t = 0$ may correspond to the instant when observation begun).
In symbols:

$$p_{ij}(t) = pr(Y_t = j \mid Y_0 = i), \qquad t \geq 0, \quad i,j = 0,1,2,\ldots .$$

Thus, $p_{ij}(t)$ expresses the transition pr. from state i
at $t = 0$ to state j at time t. Note that the values
i and j are arbitrary. For example $p_{i0}(t)$ is the pr.
that although initially there were i customers, at time t
there is none (i.e., the server is free at t). The life
time representing distance between the instant when a free
server becomes engaged, and the instant when the server is
free again, is known as a busy period. Clearly busy periods
alternate with slack periods.

Assumptions stated above are used to derive equations
for the transition pr.'s $p_{ij}(t)$.

Fig 22.1
Waiting line with a single server.

 For this purpose consider two contiguous intervals
$(0,t)$ and $(t,t+h)$ where $h > 0$ is small. Suppose that
at $t = 0$ there are already i customers present. In
order to have j customers at $t + h$, one must consider
four possibilities:

(a)   There are $j - 1$ customers at t, and one arrival
      occurred during $(t,t+h)$,

(b)   There are $j + 1$ customers at t, and one termination
      occurred during $(t,t+h)$

(c)   There are j customers at t, and no change occurred
      during $(t,t+h)$,

(d)   more than one change occurred during $(t,t+h)$.

     The arrival of a new customer means that the interar-
rival period terminates during $(t,t+h)$ and we have seen
in Section 3 that for the exponential life time this occurs
with pr. λh, approximately; indeed we have here the Poisson
input discussed in Section 19. Clearly, the arrival of a
new customer causes an increase in the value of Y_t by 1,
no matter how many customers are already in the system at
the instant of arrival. Thus, the pr. of contingency (a)
is approximately:

$$P_{j-1,j}(h) = \lambda h.$$

Similarly, the termination during $(t, t+h)$ of a service time in progress occurs with pr. μh, approximately, because service times are exponential. Clearly the termination of service causes a decrease in the value of Y_t by 1, and this is the only way customers can depart. Thus the conditional pr. of contigency (b) is approximately:

$$P_{j+1,j}(h) = \mu h.$$

By assumptions (vi) - (viii), the pr. of more than one transition during $(t, t+h)$ is negligible. Hence, the pr. of contingency (c) is approximately

$$P_{jj}(h) = 1 - \lambda h - \mu h.$$

In terms of pr.'s, the above contingencies can be expressed by the "balance equation":

$$P_{ij}(t+h) = P_{i,j-1}(t) P_{j-1,j}(h) + P_{i,j+1}(t) P_{j+1,j}(h)$$

$$+ P_{ij}(t) P_{jj}(h) + \dots .$$

Substituting the expressions found above, the balance equation becomes (up to approximation):

$$P_{ij}(t+h) = P_{i,j-1}(t) \lambda h + P_{ij}(t)(1 - \lambda h - \mu h)$$

$$+ P_{i,j+1}(t) \mu h + \dots \quad \text{for small} \quad h.$$

Hence

$$\frac{p_{ij}(t+h) - p_{ij}(t)}{h} = p_{i,j-1}(t)\lambda - p_{ij}(t)(\lambda+\mu)$$

$$+ p_{i,j+1}(t)\mu + \dots .$$

Passing to the limit as $h \to 0$, one obtains a difference-differential equation:

$$\frac{dp_{ij}(t)}{dt} = \lambda p_{i,j-1}(t) - (\lambda+\mu)p_{ij}(t) + \mu p_{i,j+1}(t) \qquad (*)$$

which is valid for all i and for $j = 1,2,\dots,$ and for all $t \geq 0$. For $j = 0$, terminations are impossible, so by the similar reasoning

$$p_{i0}(t+h) = p_{i0}(t)(1 - \lambda h) + p_{i1}(t)\mu h + \dots$$

which leads to:

$$\frac{dp_{i0}}{dt}(t) = -\lambda p_{i0}(t) + \mu p_{i1}(t) . \qquad (**)$$

Note that i and j assume all values $0,1,2,\dots$ and it is immaterial whether i is larger or smaller than j. Observe also that assumption (iii) did not intervene. It may be of interest to note that in the special case when $\mu = 0$, the above equations reduce to those of the Poisson process discussed in Section 19. It is the presence of terminations which results in the fluctuation of values of Y_t, and not merely increasing as in the Poisson case.

The initial conditions are expressed by:

$$\boxed{p_{ii}(0) = 1, \quad p_{ij}(0) = 0 \quad \text{for} \quad i \neq j} \,. \qquad (***)$$

Although the above infinite system of equations can be solved explicitly, the solution is rather complicated and therefore not very informative. For practical applications is of greater interest to see what is the solution in equilibrium, that is after a sufficiently long time when "steady state" conditions prevail. The important fact is that such a solution exists, and is independent of initial conditions (that is the effect of initial conditions wears out as time progresses), provided a rather mild and obvious restriction is imposed. Moreover, this equilibrium solution has very simple and familiar form.

Thus, we shall assume that the following limit exists for every $j = 0,1,\ldots$

$$\lim_{t \to \infty} p_{ij}(t) = P(j)$$

and is independent of the initial state i, and that P is proper distribution:

$$\sum_{j=0}^{\infty} P(j) = 1.$$

It is intuitively clear that as $p_{ij}(t)$ tends to a constant, its derivative would tend to 0:

$$\lim_{t \to \infty} \frac{dp_{ij}}{dt}(t) = 0.$$

Passing to the limit with $t \to \infty$ in equations (*) and (**), one obtains the so called "equilibrium equations":

$$
\begin{array}{ll}
0 = -\lambda P(0) + \mu P(1), & \text{for } j = 0 \\[2mm]
0 = \lambda P(j-1) - (\lambda+\mu)P(j) + \mu P(j+1), & \text{for } j = 1,2,\ldots
\end{array}
$$

One can solve this system very easily. Consider first equations for $j = 0$ and $j = 1$:

$$0 = \lambda P(0) - (\lambda+\mu)P(1) + \mu P(2) \qquad (\text{for } j = 1).$$

Substituting from equation for $j = 0$, one has from the above that

$$0 = -\lambda P(1) + \mu P(2).$$

Proceeding in the same manner for $j = 2,3,\ldots$, one reduces the above system to a much simpler form:

$$0 = -\lambda P(j) + \mu P(j+1) \qquad \text{for } j = 0,1,\ldots$$

or

$$P(j+1) = (\lambda/\mu)P(j) = \rho P(j) \qquad \text{for } j = 0,1,\ldots$$

where $\rho = \lambda/\mu$ is known as the queueing intensity. Note that ρ is the ratio of the average service time and the average interarrival time (see Example 12.4).

Hence:

$$P(1) = \rho P(0)$$
$$P(2) = \rho P(1) = \rho^2 P(0)$$

so by induction

$$P(j) = \rho^j P(0) \qquad \text{for} \quad j = 0,1,\ldots .$$

This is almost what we want, except that $P(0)$ is unknown. To find $P(0)$ we must use the fact that all $P(j)$ add to 1, so

$$1 = \sum_{j=0}^{\infty} P(j) = P(0) \sum_{j=0}^{\infty} \rho^i = P(0)(1-\rho)^{-1}$$

provided $\rho < 1$, because otherwise the geometric series would not converge. This is the required condition for existence of the steady state solution $P(j)$. It now assumes the final form

$$\boxed{P(j) = (1-\rho)\rho^j \qquad \text{for} \quad j = 0,1,\ldots} \quad .$$

This is the geometric distribution (see Section 13). Clearly $E(Y_t) = \rho/(1-\rho)$.

Section 23: Waiting Time

We shall now continue the discussion of the queueing
system described in the preceding section. We have already
found the distribution of the number of customers in the
system at an arbitrary instant t in equilibrium, that is,

$$P(j) \quad = \quad pr(Y_t = j).$$

This is the geometric distribution with parameter ρ.

We shall now consider the waiting time W for an
arbitrary customer, and we shall determine its distribution:

$$W(t) \quad = \quad pr(W \leq t), \qquad t \geq 0.$$

That is, W(t) is the d.f. of the life time W.

In order to achieve this, it is most convenient to
follow an individual customer arriving at some instant. If
the server is free, this customer is given service immediately, and his waiting time is zero, with positive pr. We
have here a new phenomenon which we did not see earlier,
namely the possibility of a life time W assuming a single
value (zero) with pr. not equal to zero.

If the server is busy, the customer joins the queue and
waits until his turn comes. Assumption (iii) on strict order
of service is essential here. Consider now

$$G_n^c(t) \quad = \quad \text{the pr. that a customer who on his arrival}$$
$$\text{has met} \quad n-1 \quad \text{waiting customers and became}$$

the n-th in the queue, has not yet received
service up to time t (for n = 1,2,...
and t ≥ 0).

This is in fact the complementary d.f. of the waiting time,
conditional that at the instant of arrival there will be
exactly n customers in the system, excluding that new
arrival. Hence, necessarily n = 1,2,... .

It is intuitively clear, and it may be verified
analytically, that the pr. of having exactly n customers
at instants of arrival should coincide with the pr. of
exactly n customers at an arbitrary instant in equilibrium.
That is P(n), found above. Hence, taking into account the
values of n (see Section 3 on conditioning), one obtains
the formula for the complementary d.f. of the waiting time:

$$W^C(t) = \sum_{n=1}^{\infty} P(n)G_n^C(t) = (1-\rho) \sum_{n=1}^{\infty} \rho^n G_n^C(t).$$

This is the pr. that the waiting time is larger than t.

It is now required to find functions G_n^C. Consider
again two intervals (0,t) and (t,t+h) where h > 0 is
small. Observe that by assumption of the strict order of
service, customers who arrive after our customer, do not
influence his waiting. Thus, during (t,t+h) only two kinds
of transitions can occur:

(a) termination of the actual service, with pr. μh (approximately), so there will be n - 1 customers in the system

(b)   no change in the number of customers (who may influence
      the waiting time); this occurs with pr. $1 - \mu h$,   assuming
      that more than one transition has negligible pr.

Thus for $n > 1$,   one has the balance equation

$$G_n^C(t+h) \;=\; \mu h G_{n-1}^C(t) \;+\; (1-\mu h) G_n^C(t)$$

whereas for $n = 1$ the balance equation is simply:

$$G_1^C(t+h) \;=\; (1 - \mu h) G_1^C(t).$$

Hence, as before, one obtains the difference-differential
equations for G_n^C:

$$
\boxed{
\begin{aligned}
\frac{dG_1^C}{dt}(t) &= -\mu G_1^C(t),\\[2ex]
\frac{dG_n^C}{dt}(t) &= \mu G_{n-1}^C(t) - \mu G_n^C(t), \qquad \text{for } n = 2,3,\ldots
\end{aligned}
}
$$

and the initial condition is clearly $G_n^C(0) = 1$.

    We must now solve this system.   It is a rather curious
fact that we do not need this explicit form of G_n^C.   Indeed,
looking at the formula for $W^C(t)$,   we only need a function
of the form:

$$\varphi(z,t) \;=\; \sum_{n=1}^{\infty} z^n G_n^C(t), \qquad \text{for } 0 \le z < 1$$

because then:

$$W^c(t) \; = \; (1-\rho)\varphi(\rho,t).$$

Now, multiply each equation for G_n^c by z^n, and then add all equations for all n. One then has

$$\frac{d\varphi}{dt} \; = \; \mu z \varphi - \mu \varphi \; = \; -(1-z)\mu \varphi$$

which is a single differential equation. The initial condition is $\varphi(z,0) = \frac{z}{1-z}$. To solve it write

$$\frac{d\varphi}{\varphi} \; = \; -(1-z)\mu \; dt.$$

Integrating one has

$$\varphi \; = \; C e^{-(1-z)\mu t}.$$

From initial conditions, the constant is found to be $C = \frac{z}{1-z}$. Hence, the complete solution is:

$$\varphi(z,t) \; = \; z(1-z)^{-1}e^{-(1-z)\mu t}.$$

(We have seen an analogous procedure in Section 20.) Hence, substituting $z = \rho$, we get our final solution:

$$W^c(t) \; = \; \rho e^{-(1-\rho)\mu t}, \qquad \text{for} \quad t \geq 0.$$

This is the required pr. that the waiting time will be longer
than t. Remarkably simple formula for so much work!
Observe that for t = 0, it gives ρ,   the pr. that the
customer need not wait (i.e., that his waiting time is zero)
is

$$W(0) = 1 - \rho.$$

This is a remarkable result. Recall that we must have
$\rho < 1$.

****

Remark: It is of interest to mention the alternative way of
solution which also leans on other problems we discussed
earlier.

    Consider again the arriving customer who joined the line
and became the n-th in the queue. There are n-1 customers
in front of him in the line, and one customer being served.
Total n life times -- service times, which must terminate
before our customer is admitted to the server. Thus his
waiting time may be represented as the sum:

$$W_n = X_1 + \ldots + X_{n-1} + X_n$$

By our assumptions, the first n - 1 life times are exponential with parameter μ; the last one is a portion of a service time since the instant of arrival of our customer until
the termination of a service (of a customer being served

when our customer joined the line). By considerations from Section 4, this remaining life time is also an exponential with the same μ. Thus, all life times forming the waiting time W_n are i.i.d.

Thus, we have here the situation discussed in Section 16 (with W_n corresponding to S_n), so the density of W_n is

$$g_n(t) = \frac{(\mu t)^{n-1}}{(n-1)!} e^{-\mu t} \mu, \qquad n = 1, 2, \ldots, \quad t \geq 0.$$

Hence, the explicit form of the complementary d.f. of W_n is:

$$G_n^c(t) = \sum_{v=0}^{n-1} \frac{(\mu t)^v}{v!} e^{-\mu t}.$$

This is the explicit form of the solution of the difference-differential equations for G_n^c. (Constrast this with the analogous equations for the Poisson distribution in Problem 2).

Substituting this expression for $G_n^c(t)$ into the formula for $W^c(t)$, one has:

$$W^c(t) = (1-\rho) \sum_{n=1}^{\infty} \rho^n \sum_{v=0}^{n-1} \frac{(\mu t)^v}{v!} e^{-\mu t}.$$

Interchanging the order of summation, one obtains the same expression for $W^c(t)$ found above.

****

Return now to the distribution of waiting time. It has the form:

$$W(t) = 1 - \rho e^{-(1-\rho)\mu t}, \qquad t \geq 0.$$

Recall that this is the pr. that the waiting time is t or less.

Differentiation would yield

$$w(t) = \rho(1-\rho)\mu e^{-(1-\rho)\mu t}$$

which however is not the density, because integrating back one only gets:

$$\int_0^t w(s)\ ds = \int_0^t \rho(1-\rho)\mu e^{-(1-\rho)\mu s}\ ds = \rho[1 - e^{-(1-\rho)\mu t}].$$

Strangely enough, this does NOT equal W(t). This effect is due to the discontinuity of W(t) at t = 0, so there is no derivative at t = 0.

Fortunately, the mean waiting time can be computed easily from the formula:

$$E(W) = \int_0^\infty W^c(t)\ dt = \rho \int_0^\infty e^{-(1-\rho)\mu t}\ dt = \rho(1-\rho)^{-1}\mu^{-1}.$$

It is convenient to compare the ratio of the mean waiting
time $E(W)$ and the mean service time $E(X) = 1/\mu$:

$$\frac{E(W)}{E(X)} = \frac{\rho}{1-\rho} .$$

This ratio increases with ρ, and equals 1 for $\rho = \frac{1}{2}$.
Note that this ratio is equal to the average number of
customers in the system, $E(Y_t)$.

<u>Remark</u>: For still another approach to the waiting time see
Problem 28.

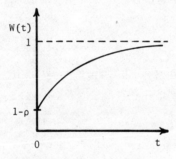

Fig 23.1

Waiting time distribution.

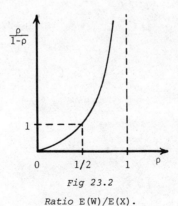

Fig 23.2

Ratio $E(W)/E(X)$.

Section 24:   Birth Right

Consider a population of material objects or living
organisms, and suppose that we are only interested in observ-
ing the increase of the size of this population (that is
"births" in the population).   For example, we may observe
a growth of a colony of bacteria, or of a herd of sheep; we
may investigate the spread of epidemics, arrivals of customers, accumulation of claims, orders and bills, etc.   The
situation is actually the same as that of input discussed
in Section 19, except that now we shall impose more general
requirements (of which the Poisson input will be a specific
case).   Basically, we shall consider the situation when the
rate of births (input) depends on the size of this population.   Our analysis will be similar to that in Section 22
(but we disregard "deaths," and generalize the input).

We shall be interested in the number of births occurring within the time interval of length   t   (and we shall
assume as before that it is the length of the interval which
matters, but not its placement on the time axis).   Let Y_t
be a r.v. representing a number of births until time   t.
Here Y_0 is the initial number of births (at the instant of
starting observations).   Clearly, Y_t assumes values
$0, 1, 2, \ldots$,   and can only increase as time progresses.   Hence
$(Y_t = j)$ is the event that at time   t   exactly   j   births
have been observed (the system is said then to be in state j).

We shall try to determine the transition pr.'s:

$$p_{ij}(t) \;=\; pr(Y_t = j \mid Y_0 = i), \quad 0 \le t$$

that at time t the system is in state j (j births
occurred), given that initially it was in state i (the
initial i births). We see immediately that $p_{ij}(t) = 0$
for $i > j$, so only the case of $j \geq i$ is of interest.

The family $(Y_t, 0 \leq t < \infty)$ is called a <u>birth process</u>,
provided the following assumptions hold:

A1: the conditional pr. that during an interval (t,t+h)
     where $h > 0$ is small, a birth occurs, given that
     at the beginning of the interval the system was in
     state i, is approximately equal to $\lambda_i h$.

A2: the conditional pr. of more than one birth during
     (t,t+h) is negligible.

Physically this implies that the number of births can
only increase in unit steps, with the pr. of a new birth independent of time but dependent on the size of the population,
and proportional to the interval length h. The factor of
proportionality λ_i is called the intensity or <u>birth rate</u>.

In order to derive equations for transition pr.'s,
consider two contiguous intervals (0,t) and (t,t+h) where
$h > 0$ is small. Suppose that at time t = 0 there are
already i births recorded. In order to have j births
at time t + h, one must consider three possibilities:

(a) there have been j - 1 births at t, and one birth
     occurred during (t,t+h),

(b)   there have been j births at t, and no birth
      occurred during (t,t+h),

(c)   more than one birth occurred during (t,t+h).

With the state j - 1 prevailing at time t, the conditional
pr. of a new birth during h depends on j - 1 and h, so
the conditional pr. of contingency a) is approximately:

$$p_{j-1,j}(h) \;=\; \lambda_{j-1}h.$$

On the other hand, the conditional pr. of contingency (b) is
approximately

$$p_{jj}(h) \;=\; 1 - \lambda_j h$$

because $\lambda_j h$ is the pr. of one birth. By A2, the pr. of
contingency (c) is negligible. In terms of pr.'s, the above
contingencies can be expressed by the "balance equation":

$$p_{ij}(t+h) \;=\; p_{i,j-1}(t)p_{j-1,j}(h) + p_{ij}(t)p_{jj}(h) + \cdots .$$

Substituting the expressions found above, the balance equation becomes (up to approximation) for small h:

$$p_{ij}(t+h) \;=\; p_{i,j-1}(t)\lambda_{j-1}h + p_{ij}(t)(1-\lambda_j h) + \cdots .$$

Hence:

$$\frac{p_{ij}(t+h) - p_{ij}(t)}{h} \;=\; p_{i,j-1}(t)\,\lambda_{j-1} - p_{ij}(t)\lambda_j + \cdots .$$

Passing to the limit as $h \to 0$, one obtains a difference-differential equation:

$$\frac{dp_{ij}}{dt}(t) = p_{i,j-1}(t)\lambda_{j-1} - p_{ij}(t)\lambda_j \qquad (*)$$

which is valid for all i and j such that $j > i$ and $i = 0,1,2,\ldots,$ and for all $t \geq 0$.

For $j = i$, only the second contingency can occur, so by similar reasoning:

$$p_{ii}(t+h) = p_{ii}(t)(1-\lambda_i h) + \ldots$$

which leads to:

$$\frac{dp_{ii}}{dt}(t) = -p_{ii}(t)\lambda_i . \qquad (**)$$

The initial conditions are expressed by:

$$p_{ii}(0) = 1, \quad p_{ij}(0) = 0 \qquad \text{for } j > i. \qquad (***)$$

Clearly, these equations are similar to equations stated in Section 22, and reduce to equations for the Poisson Process when all λ_i are equal. In contrast, with equations in Section 22, the birth equations derived here, can be easily solved.

We shall now find this solution and examine its properties. For this purpose we shall use Laplace transforms
defined in Section 7. Write:

$$u_{ij}(\alpha) \; = \; \int_0^\infty e^{-\alpha t} p_{ij}(t) \; dt, \qquad \alpha > 0.$$

We shall proceed by induction. First, multiply both sides
of equation (**) by $e^{-\alpha t}$ and integrate. Remembering the
initial condition 1, one finds (as noted in Section 7) that

$$\alpha u_{ii}(\alpha) - 1 \; = \; -u_{ii}(\alpha)\lambda_i$$

or

$$u_{ii}(\alpha) \; = \; \frac{1}{\alpha+\lambda_i}$$

which yields the explicit solution

$$p_{ii}(t) \; = \; e^{-\lambda_i t}.$$

Taking the transform of equation (*), and remembering that
the initial conditions vanish for $i \neq j$ one has similarly:

$$\alpha u_{ij}(\alpha) \; = \; u_{i,j-1}(\alpha)\lambda_{j-1} - u_{ij}(\alpha)\lambda_j$$

or

$$u_{ij}(\alpha) \; = \; u_{i,j-1}(\alpha) \frac{\lambda_{j-1}}{\alpha+\lambda_j}, \qquad \text{for} \quad j > i.$$

As the product of Laplace transforms is the transform of convolution, the above relation becomes

$$p_{ij}(t) = \lambda_{j-1} \int_0^t e^{-\lambda_j(t-s)} p_{i,j-1}(s)\, ds, \qquad j > i.$$

This may be used to calculate $p_{ij}(t)$ recursively. It is more convenient, however, to work with Laplace transforms. The above recurrence relations for $u_{ij}(\alpha)$ can be solved directly (see a similar argument in Section 22) to give:

$$u_{ij}(\alpha) = \frac{1}{\alpha+\lambda_j} \frac{\lambda_{j-1}}{\alpha+\lambda_{j-1}} \cdots \frac{\lambda_i}{\alpha+\lambda_i} \qquad \text{for } j > i.$$

And this is our final solution. All we need do, is to substitute for the coefficients $\lambda_i,\ldots,\lambda_j$ and to invert the Laplace transform to get $p_{ij}(t)$. For example, if all $\lambda_i = \lambda$, then

$$u_{ij}(\alpha) = \frac{1}{\alpha+\lambda} \left(\frac{\lambda}{\alpha+\lambda}\right)^{j-i}, \qquad j \geq i$$

which is the Laplace transform of the Poisson distribution.

However, a lot of useful information can be obtained directly from the Laplace transform, without finding $p_{ij}(t)$ explicitly. For example, recall from Section 7 that

$$\lim_{t\to\infty} p_{ij}(t) = \lim_{\alpha\to0} \alpha u_{ij}(\alpha)$$

and it follows from the above expression for $u_{ij}(\alpha)$ that for all i and j:

$$\lim_{t \to \infty} p_{ij}(t) = 0$$

(in contrast with the limiting distribution in Section 22).
This means that as time increases, the pr. of having any
finite number of births goes to zero (so we may expect that
the number of births will increase indefinitely).

Persuing this observation further, consider the expres-
sion $u_{ij}(\alpha)\lambda_j$ and note that it is the product of Laplace
transforms of exponential distributions. In particular, the
exponential form of $p_{ii}(t)$ means that the length of time
T_i the system spends in state i is exponentially distributed with the parameter λ_i. It then follows that the life
time T_j (i.e., time spent in state j) has exponential
distribution with the parameter λ_j, hence mean $1/\lambda_j$, for
each $j \geq i$:

$$pr(T_j > t) = e^{-\lambda_j t}.$$

Consider now the total time T spent in all states --
the life time of the birth process -- defined by

$$T = \sum_{j=i}^{\infty} T_j.$$

Taking expectations, the average total life time is clearly

$$E(T) = \sum_{j=1}^{\infty} \frac{1}{\lambda_j}.$$

Intuitively, we would expect $E(T)$ to be infinite, like in

the Poisson process. It is strange, however, that $E(T)$ may be finite in some cases (depending on the choice of λ_j).

Indeed, T itself is a rather strange life time (it is a duration of the whole birth process) and it is natural to expect that it will be infinite (and then obviously $E(T)$ would be infinite, too). Well, not quite! Let us write

$$F^c(t) = pr(T > t \mid Y_0 = i)$$

for the (complementary) d.f. of T, conditional on the initial state i (we shall drop i from the notation for simplicity). Taking the limit as $t \to \infty$, $F^c(\infty)$ is the pr. that T is infinite, and $F(\infty)$ is the pr. that T is finite.

It can be shown (by a renewal argument similar to that in Section 16) that

$$F^c(t) = \sum_{j=i}^{\infty} P_{ij}(t).$$

In view of the above remarks, as well as by the total probability argument, we would expect the sum to be equal to 1. This would correspond to T being infinite with pr. 1, i.e., $F(t) = 0$ for all finite t and $F(\infty) = 0$ or $F^c(\infty) = 1$.

But curious thing happen! There are cases (depending on the choice of λ_j) that

$$\sum_{j=i}^{\infty} P_{ij}(t) \leq 1$$

so $F(t)$ need not be zero for finite t, and consequently

F(t) represents the distribution of T, with F(∞) being
the pr. that T is finite.

But T is the total time for the process to reach
infinitely many births, so T finite means that infinitely
many births will occur in finite time. Some kind of explosion! It can be shown (by analytic methods which we cannot
discuss here) that F(∞) can be either 0 or 1 only, and
that:

$F^c(\infty) = 1$ (T infinite with pr. 1) if and only if $E(T) = \infty$

$F^c(\infty) = 0$ (T finite with pr. 1)   if and only if $E(T) < \infty$.

This is a very interesting mathematical result. We already
noted that for the Poisson input T is infinite with pr. 1
and the series for E(T) obviously diverges.

Similarly, if we take $\lambda_j = j\lambda$ for $j \geq 1$, and remember
that

$$\sum_{j=1}^{\infty} \frac{1}{j} = \infty,$$

then again $E(T) = \infty$ (for every i), so T is infinite
with pr. 1.

On the other hand, taking $\lambda_j = j^2\lambda$ for $j \geq 1$ and
using the fact that

$$\sum_{j=1}^{\infty} \frac{1}{j^2} = \frac{\pi^2}{6} < \infty,$$

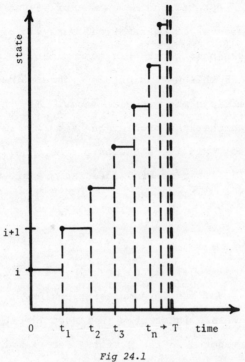

Fig 24.1

Birth process.

one finds that $E(T) < \infty$, so T is finite with pr. 1.
Thus, the explosion does occur, indeed.

On this thoughtful note, we close our discussion of
life times.

Section 25: Birth and Death

A dilligent reader of this chapter has noticed no doubt
a common pattern in our discussion of transition probabilities
$p_{ij}(t)$. Indeed, it is really cumbersome to repeat each time

a problem is presented the same line of reasoning. Fortunately the character of mathematical analysis offers an economy
of thought by singling out this common pattern and study its
properties. In this way we obtain a general model of importance which embraces many special cases. All that is needed
is to check whether the problem on hand fits into our general
model -- if so, then the solution of the model will automatically hold in our special case, and we are spared lengthy
derivations.

25.1.

Our general model may be described as follows. We have
a certain physical system comprising equipment, people or
animals, and we are interested in random fluctuations of a
number of components in the system. For example, we may look
at the number of operating machines in a work shop, the number of patients in the waiting room, the number of fish in
a pond, the number of students in a class, etc. To be more
definite, we shall say that we wish to investigate the number of customers in the system. But we shall leave to the
reader's imagination what is meant by a system and what is
meant by a customer (if you are still at a loss, think about
the queueing system in Section 22). Clearly, the number of
customers is 0,1,..., etc. until say, we allow a finite
number (or the number may be infinite). We shall say that
the system is in the state i when there are exactly i
customers present. In our picturesque language, we shall
regard the increase in the number of customers as a birth,

and decrease as a death. Hence the name "birth and death" process.

In probabilistic formulation, we shall consider a family of r.v.'s (discrete life times) Y_t for $0 \le t < \infty$, where each Y_t represents the state of the system at the instant t. Thus, $(Y_t = i)$ is the event that at time t, there are exactly i customers in the system (i.e., the system is in state i). We shall consider transition probabilities:

$$p_{ij}(t) = pr(Y_t = j \mid Y_0 = i) \quad \text{for } i,j = 0,1,\ldots, \quad t \ge 0$$

that at time t the system is in state j, given that it was initially in state i. We shall impose an assumption of homogeneity which means that this pr. $p_{ij}(t)$ is the same for any pair Y_{t+s} and Y_s, irrespective of s.

We shall now impose a rather powerful assumption (which we tacitly used in this chapter) that our process has the Markov property, namely that the transition pr.'s satisfy the "balance equation":

$$p_{ij}(t+h) = \sum_k p_{ik}(t)p_{kj}(h), \quad t \ge 0, \quad h \ge 0.$$

The meaning of this equation can be explained as follows. Suppose that at time 0 the system is in state i, but at time $t+h$ it should be in state j. How does it go from i to j ? First it moves to state k at time t, and then from k to j during the remaining time h. The asso ciated pr.'s are $p_{ik}(t)$ and $p_{kj}(h)$, respectively, and their product $p_{ik}(t)p_{kj}(h)$ is the pr. of the path from i through k to j. Summation over all k (all paths) yields the required pr. of transition. The Markov property is expressed by the fact that $p_{kj}(h)$ is independent of the state i at time 0. In other words, the transition from a state to another state depends on these states only, and not on the way the first state has been reached.

We also have the obvious requirements:

$$p_{ij}(t) \geq 0, \quad \sum_j p_{ij}(t) = 1, \quad \lim_{t \to 0} p_{ij}(t) = \begin{cases} 0 & \text{if } i \neq j \\ 1 & \text{if } i = j \end{cases} .$$

The family $(Y_t, \ 0 \leq t < \infty)$ is called a <u>birth</u> <u>and</u> <u>death</u> <u>process</u> if the following limits exist for each j:

$$\lim_{h \to 0} \frac{p_{j,j+1}(h)}{h} = \lambda_j, \qquad \qquad \lim_{h \to 0} \frac{p_{j,j-1}(h)}{h} = \mu_j,$$

$$\lim_{h \to 0} \frac{p_{jr}(h)}{h} = 0 \qquad \text{for } r \neq j-1, \ j, \ j+1$$

and

$$\lim_{h \to 0} \frac{1 - p_{jj}(h)}{h} = \lambda_j + \mu_j .$$

Probabilistically, these assumptions mean that:

A1: the conditional pr. that during an interval (t,t+h) where h > 0 is small, a transition (birth) to state j + 1 occurs, given that at t the system was in state

j, is

$$p_{j,j+1}(h) = \lambda_j h \quad \text{approximately;}$$

A2: the conditional pr. that during an interval (t,t+h)

where h > 0 is small, a transition (death) to state

j - 1 occurs, given that at t the system was in state

j, is

$$p_{j,j-1}(h) = \mu_j h \quad \text{approximately;}$$

A3: the conditional pr. of more than one transition during

(t,t+h) is negligible, and of no transition is

$$1 - p_{jj}(h) = (\lambda_j + \mu_j)h \quad \text{approximately.}$$

Physically, this implies that the system changes (during
h) from a state to its next neighbor only, or no change takes
place. Thus we have births and deaths by a unit only, the
multiple transitions being excluded; see Fig. 25.1. The
constants λ_j and μ_j are called the birth and death
intensities or <u>rates</u>. Clearly $\mu_0 = 0$ and if the number of
states is finite, say n, then $\lambda_n = 0$.

In order to derive equations for the transition pr.'s,
we shall use our balance equation for $p_{ij}(t+h)$. Separating
the three terms corresponding to k = j - 1, j, j + 1, we
have:

$$p_{ij}(t+h) = p_{i,j-1}(t)p_{j-1,j}(h) + p_{ij}(t)p_{jj}(h)$$

$$+ p_{i,j+1}(t)p_{j+1,j}(h) + \sum p_{ik}(t)p_{kj}(h)$$

hence:

$$\frac{p_{ij}(t+h) - p_{ij}(t)}{h} = p_{i,j-1}(t)\frac{p_{j-1,j}(h)}{h} - p_{ij}(t)\frac{1 - p_{jj}(h)}{h}$$

$$+ p_{i,j+1}(t)\frac{p_{j+1,j}(h)}{h} + \sum p_{ik}(t)\frac{p_{kj}(h)}{h}$$

where the above sums are extended over $k \neq j-1$, j, $j+1$.

Passing to the limit with $h \to 0$, we obtain the basic birth and death equation:

$$\boxed{\frac{dp_{ij}}{dt}(t) = p_{i,j-1}(t)\lambda_{j-1} - p_{ij}(t)(\lambda_j + \mu_j) + p_{i,j+1}(t)\mu_{j+1}} \qquad (*)$$

for $i \geq 0$ and $j > 0$. For $j = 0$, no death can occur, so the same reasoning yields:

$$\frac{dp_{i0}}{dt}(t) = -p_{i0}(t)\lambda_0 + p_{i1}(t)\mu_1,$$

and if the number of states is finite, then (no birth possible) the last equation is :

$$\frac{dp_{in}}{dt}(t) = p_{i,n-1}(t)\lambda_{n-1} - p_{in}(t)\mu_n.$$

The initial conditions are expressed by:

$$p_{ii}(0) = 1, \qquad p_{ij}(0) = 0 \qquad \text{for } i \neq j.$$

It is now clear that our equation (*) is the generalization of equations encountered earlier. When all $\lambda_j = \lambda$ and all $\mu_j = 0$, we get the Poisson process, and when $\mu_j = 0$ for all j we obtain the birth process (Section 24). In Section 20 there are only two states (n = 1) and $\lambda_0 = b$, $\mu_0 = 0$, $\lambda_1 = 0$, $\mu_1 = a$. Finally, in Section 22 we have $\lambda_j = \lambda$ and $\mu_j = \mu$. It is also clear that our considerations of all possible contingencies in those sections are now embraced by our single derivation here from the balance equation.

Now, for the practical applications, all we need to do is to specify the rates λ_j and μ_j, and to substitute them in the ready-made single equation (*). You see the saving of labor achieved in this manner! The choice of coefficients λ_j and μ_j depends of course on the problem on hand. Earlier sections in this chapter provide ample examples, but we shall work out a few more soon. Incidently, note that we considered transitions during h, following the instant t -- for this reason equation (*) is conveniently called the forward equation.

25.2.

Although we got our equation (actually the infinite system of difference-differential equations), the solution is hard to find, and even if we get one it may be so complicated that it is practically useless. And besides this, strange things may happen as we have seen in the case of the birth process (in Section 24).

Therefore, it is of great practical interest to know if there exists the equilibrium solution, that is the solution which may be reached after a sufficiently long time. We found such a solution in the queueing problem in Section 22 and also in the learning model in Section 20, but we have seen that there is no such a solution for the birth process.

Thus, we shall assume that the following limit exists for every j and is independent of the initial state i:

$$\lim_{t \to \infty} p_{ij}(t) = P(j)$$

and that P is a proper distribution:

$$\sum_j P(j) = 1.$$

It is intuitively clear that then $dp_{ij}(t)/dt \to 0$. Passing to the limit with $t \to \infty$ in equation (*), one obtains the so called equilibrium or steady-state equations:

$$
\begin{array}{ll}
-\lambda_0 P(0) + \mu_1 P(1) = 0 & j = 0 \\
\\
\lambda_{j-1} P(j-1) - (\lambda_j + \mu_j) P(j) + \mu_{j+1} P(j+1) = 0 & j > 0
\end{array}
$$

and if the number of states is finite, the last equation is:

$$\lambda_{n-1} P(n-1) - \mu_n P(n) = 0, \quad j = n.$$

These equations are very common in practical applications, and may be solved explicitly. Proceeding as in Section 22

(by successive elimination) we can express the solution in
the form:

$$P(j) \;=\; P(0)\; \frac{\lambda_0 \lambda_1 \cdots \lambda_{j-1}}{\mu_1 \mu_2 \cdots \mu_j}$$

where $P(0)$ must be determined from the fact that all $P(j)$
add to 1. It is easy to check that solutions obtained in
Sections 20 and 22 can be easily obtained from this formula.
If you think that we pulled your leg by going through lengthy
calculations in these sections to obtain the result which we
now got effortlessly, rest assured that we have been far from
such evil thoughts. We simply chose this way to show you the
beauty of the mathematical models of general form. In this
vein we may add that the birth and death process is itself a
special case of the more general Markovian models.

Example 1: (Telephone exchange). The number of lines in a
telephone exchange is large, and may be assumed infinite.
Let Y_t be the number of busy lines at time t. Fluctuations
are caused by arrivals of calls (births) and by termination
of conversations (deaths). We shall assume that the input
is Poissonian and that deaths are proportional to the number
of busy lines. Translating this description into birth and
death rates, we have for all $0 \le j < \infty$:

$$\lambda_j = \lambda, \qquad \mu_j = j\mu.$$

Equilibrium equations can be easily written down, and the
solution is found to be:

$$P(j) = \frac{A^j}{j!} e^{-A}, \qquad \text{where} \quad A = \lambda/\mu, \qquad j = 0,1,\ldots$$

again Poisson!

Example 2: (Servicing of machines). Suppose that in the
work shop there are n machines and as usual they break
down from time to time. Let Y_t be the number of nonworking
machines. We shall assume that the probability of the break-
age is proportional to the number of operating machines
(births). The broken machine is put back into service and
the time needed for repair is assumed to be the exponential
life time. Thus, deaths correspond to the decrease in the
number of inoperative machines. Again, translation into our
birth and death coefficients yields:

$$\lambda_j = (n-j)\lambda, \qquad \mu_j = j\mu \qquad \text{for} \quad j = 0,1,\ldots,n$$

(we have tacitly assumed independence of the machines).

Equilibrium equations can be easily written down, and
the solution is found to be:

$$P(j) = \binom{n}{j} p^j (1-p)^{n-j}, \qquad \text{where} \quad p = \lambda(\lambda+\mu)^{-1}, \quad j = 0,1,\ldots,n,$$

again binomial! (Compare this with subsection 21.2.)

25.3

In closing our discussion, it is hard to omit mentioning
how a trivial change leads to serious consequences. If you
look again at our balance equation in subsection 25.1, then
interchanging h with t does not affect the equation at
all. Yet, it leads to very important new results. Indeed,
h is now on the left side and this means that we now consider transitions before the instant t. Proceeding exactly
in the same manner as before, we obtain the dual system of
equations for $p_{ij}(t)$ -- we leave cheerfully their derivation
to the reader:

$$\frac{dp_{ij}}{dt}(t) = \lambda_i p_{i+1,j}(t) - (\lambda_i + \mu_i) p_{ij}(t) + \mu_i p_{i-1,j}(t)$$

with the same initial conditions as before. This new equation is called the backward equation (it has nothing wrong
with it -- the name signifies only the limit operation "on
the back"). Note that in the backward equation the second
index j is fixed, whereas in the forward equation it is
the first index i which is fixed. In the forward equation
the transitions are to the state j, but in the backward
equation the transitions are from the state i. It can be
shown that in most situations (in particular when the number of states is finite) both equations have the same solution. And let us leave worrying about different solutions
of these equations to Ph.D.'s in Mathematics.

Suppose that the limit $P(j)$ exists. Then, passing to the limit as $t \to \infty$, the backward equation yields the triviality $0 = 0$. Do not, however jump to the conclusion that the backward equation is useless. Just the opposite. It is much more important than the forward equation. Indeed, all problems concerning (continuous) life times in the birth and death processes are described by the backward equations. If you look closer at equtions for the waiting time distributions $G_n^c(t)$ in Section 23, you will notice that we have there the backward equation (with $\lambda_i = 0$ and $\mu_i = \mu$). Although we hinted at this situation from time to time, we cannot enter into the fascinating study of the backward equation and shall content ourselves with a rather interesting example.

Example 3: (Linear growth). Suppose that intitially there are i customers in the system, where $i \geq 1$. Due to fluctuations, it is possible that at some instant the number of customers will drop to 0. We would like to find the probability of this extinction. There are many situations which are covered by this model, the most famous being the "extinction of family surnames." Another is the busy period in the queueing system.

We shall consider the case of linear coefficients:

$$\lambda_i = i\lambda, \qquad \mu_i = i\mu, \qquad \text{for } i = 0,1,\ldots .$$

Clearly for i = 0, both coefficients vanish and it is impossible to leave the state 0 (thus, 0 is the absorbing state). We shall assume that i > 0, and shall consider the transition pr. from i to j = 0, namely $p_{i0}(t)$. It is intuitively clear that $p_{i0}(t)$ is in fact the distribution function of the (life) time to extinction. On the other hand, $p_{00}(t) \equiv 1$.

The backward equation for j = 0 and for i > 0 is:

$$\frac{dp_{i0}}{dt}(t) = i\lambda p_{i+1,0}(t) - i(\lambda+\mu)p_{i0}(t) + i\mu p_{i-1,0}(t).$$

Differentiation will show that the solution is of the form:

$$p_{i0}(t) = (\frac{\mu}{\lambda})^{i} (\frac{e^{(\lambda-\mu)t} - 1}{e^{(\lambda-\mu)t} - \frac{\mu}{\lambda}})^{i}, \quad i = 1,2,\dots .$$

Passing to the limit with t → ∞, it is easy to see that:

$$p_{i0}(\infty) = \begin{cases} 1 & \text{for } \lambda < \mu \\ (\frac{\mu}{\lambda})^{i} & \text{for } \lambda \geq \mu \end{cases}.$$

Therefore, when λ < μ, $p_{i0}(t)$ is a proper d.f. and the pr. that the time to extinction is finite is 1. Hence the

extinction will take place. On the other hand, when $\lambda > \mu$
then the pr. of extinction is less than one, and there is a
positive pr. $1 - p_{i0}(\infty)$ that the number of customers in
the system will increase over all bounds. For $\lambda = \mu$,
extinction will take place, but the average time will be
infinite. (Does this remind you about the fate of our rabbit in Section 18; and about the explosion in Section 24 ?).

****

Thus, we return to where we started -- to talking about
life times: our main topic of discussion. And each life
begins and ends -- so it seems fit that we conclude our
investigations with a talk about birth and death.

****

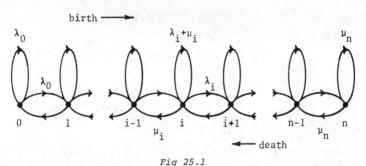

Fig 25.1

Birth and death process.

Problems: Chapter 4

1. If customers arrive at the rate of 2 per minute, what is the pr. that the number of customers arriving in 2 minutes is:

   (i) exactly 3, (ii) 3 or less, (iii) 3 or more, (iv) more than 3, (v) less than 3.

   Suppose there was initially 4 customers present. What was the average number of customers present at the end of a 2 minute period?

2. Let $p_{ij}(t)$ be the Poisson distribution from Section 19. Show that

   (i) $\int_0^\infty e^{-\alpha t} p_{ij}(t)\, dt = \dfrac{\lambda^{j-i}}{(\lambda+\alpha)^{j-i+1}}$ for $j \geq i$;

   (ii) the value of $dp_{ij}(t)/dt$ at $t = 0$ is $-\lambda$ if $i = j$, λ if $j = i+1$ and 0 otherwise;

   (iii) for fixed i and j, plot the graph of $p_{ij}(t)$ as a function of t (see problems 4 and 5, Chapt. 3).

3. Suppose that the random time T (of Section 19) has density of the form:

$$f(t) = \frac{(\mu t)^{n-1}}{(n-1)!} e^{-\mu t} \mu, \quad t > 0, \quad n = 1, 2, \ldots .$$

Show that

$$pr(N_T = j) = \binom{n+j-1}{j} \frac{\mu^n \lambda^j}{(\mu+\lambda)^{n+j}}, \quad j = 0,1,\ldots$$

and $EN_T = n\lambda/\mu$.

4. For the Poisson input (Section 19), let φ be the cost function associated with states of the form:

$$\varphi(j) = z^j \quad \text{for} \quad 0 < z < 1.$$

Show that the average conditional cost is

$$E(\varphi(Y_t) \mid Y_0 = i) = z^i e^{-\lambda t(1-z)} \le z^i.$$

5. Calculate all pr.'s in Section 20 for the following conditions:

(i) $a = b$, $\pi_0 \ne \pi_1$;

(ii) $\pi_0 b = \pi_1 a$;

(iii) $\pi_0 = \pi_1$, $a \ne b$.

Plot the corresponding graphs for selected values at a, b, and π_0, π_1.

6. Let $\varphi(i)$ be the cost function $(i = 0,1)$ associated with the learning model in Section 20. Evaluate the average cost:

$$E\varphi(Y_t) = \varphi(0)\pi_0(t) + \varphi(1)\pi_1(t).$$

Show that the average cost is:

(i)   constant if $\varphi(0) = \varphi(1)$;

(ii) $C(\pi_0 b - a\pi_1)e^{-(a+b)t}$ if $\varphi(0) = bC$, $\varphi(1) = -aC$

(where C is constant).

7. Suppose that a clever and hungry monkey tries to reach a banana with a stick. Let $a/b = k > 1$. Show that the pr. that the monkey will eventually learn to catch the banana is

$$\frac{1}{2} < \frac{k}{1+k} < 1.$$

If the monkey receives a reward $c(k+1)$ for a correct response, and $-c(k+1)$ for a wrong response, show that the average reward at time t is

$$c[k-1 + 2(\pi_0 - \pi_1 k)e^{-(1+k)bt}].$$

8. Verify that $p_{ij}(t)$ from Section 20 satisfy the integral equations in Section 21, by

(i)   direct substitution;

(ii)   comparison of Laplace transform

$$u_{ij}(\alpha) = \int_0^\infty e^{-\alpha t} p_{ij}(t)\, dt.$$

In both cases the Laplace transforms are:

$$u_{00}(\alpha) = \frac{\alpha + a}{\alpha(\alpha + a + b)} , \quad u_{01}(\alpha) = \frac{b}{\alpha(\alpha + a + b)}$$

$$u_{10}(\alpha) = \frac{a}{\alpha(\alpha + a + b)} , \quad u_{11}(\alpha) = \frac{\alpha + b}{\alpha(\alpha + a + b)} .$$

9. Differentiate expression $p_{ij}(t)$ from Section 21.2 and
    verify that it satisfies the birth and death equation
    with coefficients

    $$\lambda_j = (n-j)b, \quad \mu_j = ja \qquad (j = 0,1,\ldots,n)$$

    (see Example 2, in Section 25).

10. In the group of n lines (Section 21.2) suppose that
    there are j lines busy and n-j free. Let X_0 and
    X_1 be the idle period and the busy period for each
    line, where X_0 and X_1 have exponential distribution
    with parameters b and a, respectively.

    Define the function T of the state j to be the
    minimum of j life times X_1 and (n-j) life times
    X_0. Show that T has an exponential distribution
    with the parameter $q_j = (n-j)b + ja$ (use $G_-^c(t)$ from
    Section 6).

11. In the queueing model (Section 22), let the average
    interarrival time be 10 minutes, and the average
    time spent at the server be 5 minutes. Show that

    $$P(j) \;=\; (\tfrac{1}{2})^{j+1}, \quad j = 0,1,\ldots,$$

    and $E(Y_t) = 1$. Show also that the average waiting
    time (Section 23) is $E(W) = 5$ minutes.

12. Suppose that the cost function associated with the
    steady state distribution in the queueing model (Sec-

tion 22) is given by $\varphi(j) = z^j$, $|z| \leq 1$. Show that
the average cost is then $(1-\rho)/(1-\rho z)$.

13. Plot the graph of the mean number of customers in the
system $\rho/(1-\rho)$, as a function of ρ (for $0 < \rho < 1$).
Interpret the limits when $\rho \to 0$ and $\rho \to 1$. Show
that the mean is less than 1 when $\rho < \frac{1}{2}$; interpret.

14. For the waiting time W (Section 23) evaluate

$$EW^2 = 2 \int_0^\infty tW^c(t) \, dt,$$

and deduce that

$$\text{var } W = \frac{2\rho - \rho^2}{\rho^2(1-\rho)^2} .$$

15. For the waiting time W (Section 23) evaluate

$$\int_0^\infty e^{-\alpha t}W^c(t) \, dt,$$

and deduce (from Section 7) that the Laplace transform
of the waiting time distribution is

$$\int_0^\infty e^{-\alpha t} \, dW(t) = (1-\rho) \frac{\mu+\alpha}{\mu-\lambda+\alpha} .$$

Derive again expressions for $E(W)$ and $\text{var}(W)$.

16. In Section 23, verify that $W^c(t)/E(W)$ defines an
exponential density with mean $(\mu-\lambda)^{-1}$ (Cf. Section 4).
Interpret!

17. Show that the pr. of waiting more than the average
     waiting time is $\rho e^{-\rho}$.

18. Suppose that the cost function in the waiting time
     problem (Section 23) is linear $\varphi(t) = at + b$ (for
     $t \geq 0$). Show that the average cost of waiting is

$$E\varphi(W) \;=\; b + \frac{\rho a}{\mu(1-\rho)} \;.$$

19. In the waiting time problem (Section 23), suppose that
     the pr. of no waiting is given, say p, and that the
     average waiting time is known, say m. Show that:

$$\lambda = \frac{(1-p)^2}{mp} \;, \qquad \mu = \frac{1-p}{mp} \;.$$

20. In the birth process (Section 24), let the transition
     pr. be given for $i \geq 1$ by:

$$P_{ij}(t) \;=\; \binom{j-1}{j-i}(1-e^{-\lambda t})^{j-i} e^{-\lambda i t}, \qquad j \geq i.$$

     By differentiation, verify that $\mathrm{pr}_{ij}(t)$ satisfies
     the backward equation of the form

$$\frac{dp_{ij}}{dt}(t) \;=\; \begin{cases} \lambda_i P_{i+1,j}(t) - \lambda_i P_{ij}(t) & \text{for } i \leq j-1 \\[2mm] -\lambda_i P_{ii}(t) & \text{for } i = j \end{cases}$$

     with $\lambda_i = i\lambda$. By rearrangement of terms verify that
     the forward equation holds, too.

21. Suppose that the input stops altogether when the number of births reaches a fixed number, say n (this is called "finite input"), so

$$\lambda_j = (n-j)\lambda \quad \text{for} \quad j = 0,1,\ldots,n.$$

(i) Verify that

$$p_{ij}(t) = \begin{cases} \binom{n-i}{j-i} e^{-(n-j)\lambda t}(1-e^{-\lambda t})^{j-i}, & i \le j \le n \\ \\ 0, & i > j \end{cases}$$

is the solution of the birth equation with the above coefficient.

(ii) Show that $E(Y_t \mid Y_0 = i) = (n-i)(1-e^{-\lambda t})$,
$\text{var}(Y_t \mid Y_0 = i) = (n-i)(1-e^{-\lambda t})e^{-\lambda t}$.

22. Verify Examples 1 and 2 in Section 25.

23. Give the full derivation of the backward equation in Section 25.

24. With reference to Section 25, verify that the derivatives at $t = 0$ are:

$$\frac{dp_{i,i+1}}{dt}(t)\bigg|_{t=0} = \lambda_i, \qquad \frac{dp_{i,i-1}}{dt}(t) = \mu_i,$$

$$\frac{dp_{ii}}{dt}(t)\bigg|_{t=0} = -(\lambda_i+\mu_i).$$

25. (Death process). Show that the binomial distribution

$$p_{ij}(t) = \begin{cases} \binom{i}{j} e^{-\mu j t} (1-e^{-\mu t})^{i-j} & \text{for } 0 \leq j \leq i \\ \\ 0, & \text{for } j > i \end{cases}$$

satisfies the death equation, i.e., $\mu_j = j\mu$ for
$j = 1,2,\ldots,i$, and $\lambda_j \equiv 0$. In this problem, one
starts with i items initially, and only terminations
are allowed, the state $j = 0$ being the graveyard.

26. (Infinite number of lines). This is a generalization
of the case presented in Section 21.2 when n is
infinite. Consider the transition pr. of the form:

$$p_{ij}(t) = \sum_k r_i(j-k,t) q_i(k,t)$$

with summation over $\min(j-i,0) \leq k \leq j$, where

$$r_i(j-k,t) = \binom{i}{j-k} e^{-\mu t(j-k)} (1-e^{-\mu t})^{i-j+k}$$

and

$$q_i(k,t) = \frac{[A(t)]^k}{k!} e^{-A(t)}$$

with

$$A(t) = \frac{\lambda}{\mu} (1-e^{-\mu t}).$$

This form follows from that in Section 21.2 by passage
to the limit, is in Section 14. Verify that $p_{ij}(t)$

satisfies the birth and death equations with

$$\lambda_j = \lambda, \quad \mu_j = j\mu \quad \text{for} \quad 0 \le j < \infty.$$

Check also that the steady state solution $P(j)$ exists (as $t \to \infty$) and coincides with that given in Example 1 (Section 25).

27. Let $p_{ij}(t)$ be the transition probability of the Poisson process (Section 19).

Show that:

$$p_{ij}(t) \to 0 \quad \text{as} \quad t \to \infty$$

but (summing over all even j) one has for $t \to \infty$:

$$\sum_{j \text{ even}} p_{ij}(t) = \begin{cases} e^{-\lambda t} \cosh \lambda t \to 1/2, & \text{when } i \text{ even} \\ e^{-\lambda t} \sinh \lambda t \to 1/2, & \text{when } i \text{ odd} \end{cases}$$

28. Let M be a random variable with d.f. G_∞ obtained in Problem 19, Chapt. 3. Consider a life time M^+ obtained by truncation of M at zero:

$$M^+ = \max(0, M)$$

(see Section 6).

Show that the d.f. of M^+ coincides with the d.f. W of the waiting time obtained in Section 23. (This constitutes another approach to the study of waiting times.)

Appendix A

Formulae

Arithmetic progression:

$$a_n = a_1 + (n-1)h, \qquad a_1 + \ldots + a_n = \frac{n}{2}(a_1 + a_n)$$

$$1 + 2 + \ldots + n = \frac{1}{2}n(n+1)$$

$$1^2 + 2^2 + \ldots + n^2 = \frac{1}{6}n(n+1)(2n+1).$$

Geometric progression:

$$1 + \alpha + \ldots + \alpha^n = \frac{1 - \alpha^{n+1}}{1 - \alpha}$$

$$\sum_{i=0}^{n} i\alpha^i = \frac{\alpha}{1-\alpha}\left[\frac{1-\alpha^n}{1-\alpha} - n\alpha^n\right]$$

arithmetic mean: $\quad a = \dfrac{a_1 + \ldots + a_n}{n}$

geometric mean: $\quad g = \sqrt[n]{a_1 \ldots a_n}$

harmonic mean: $\quad \dfrac{1}{h} = \dfrac{1}{a_1} + \ldots + \dfrac{1}{a_n}$.

Series:

geometric:

$$\sum_{n=0}^{\infty} \alpha^n = \frac{1}{1-\alpha} , \quad |\alpha| < 1.$$

exponential:

$$e^x = \sum_{n=0}^{\infty} \frac{x^n}{n!} , \quad \text{all} \ x,$$

$$\sin x = \sum_{n=1}^{\infty} (-1)^{n-1} \frac{x^{2n-1}}{(2n-1)!} , \quad \cos x = \sum_{n=0}^{\infty} (-1)^n \frac{x^{2n}}{(2n)!} ,$$

$$\text{all} \ x,$$

$$\sinh x = \sum_{n-1}^{\infty} \frac{x^{2n-1}}{(2n-1)!} , \quad \cosh x = \sum_{n=0}^{\infty} \frac{x^{2n}}{(2n)!} ,$$

$$\text{all} \ x,$$

$$e^{-x^2} = \sum_{n=0}^{\infty} (-1)^n \frac{x^{2n}}{n!} , \quad \text{all} \ x.$$

logarithmic:

$$\log(1+x) = \sum_{n=1}^{\infty} (-1)^{n-1} \frac{x^n}{n} , \quad |x| < 1$$

$$\frac{1}{2} \log \frac{1+x}{1-x} = \sum_{n=1}^{\infty} \frac{x^{2n-1}}{2n-1} , \quad |x| < 1.$$

binomial expansion:

$$(a+b)^n = \sum_{k=0}^{n} \binom{n}{k} a^k b^{n-k}, \quad n = 0,1,\ldots .$$

binomial series:

$$(1+x)^{\alpha} = \sum_{n=0}^{\infty} \binom{\alpha}{n} x^{n}, \qquad \text{real}, \quad |x| < 1,$$

$$(1+x)^{-\alpha} = \sum_{n=0}^{\infty} (-1)^{n} \binom{\alpha+n-1}{n} x^{n}, \qquad \alpha > 0, \quad |x| < 1,$$

$$(1-x)^{-1} = 1 + x + x^{2} + x^{3} + \ldots,$$

$$(1-x)^{-2} = 1 + 2x + 3x^{2} + 4x^{3} + \ldots, \quad |x| < 1.$$

polynomial expansion:

$$(a_1+\ldots+a_k)^n = \sum \frac{n!}{r_1!\ldots r_k!} a_1^{r_1} \ldots a_k^{r_k}, \quad r_1+\ldots+r_k = n.$$

harmonic series:

$$1 + \frac{1}{2} + \frac{1}{3} + \frac{1}{4} + \ldots \quad \text{diverges};\qquad \sum_{n=1}^{\infty} \frac{1}{n^2} = \frac{\pi^2}{6} .$$

Binomial coefficients:

$$\binom{\alpha}{k} = \frac{\alpha(\alpha-1)\ldots(\alpha-k+1)}{k!}, \qquad \alpha \text{ real}, \quad k = 0,1,2,\ldots$$

for n positive integer or zero: $\binom{n}{k} = \dfrac{n!}{k!(n-k)!}$

$$\binom{n}{k} = \binom{n}{n-k}, \qquad \binom{n}{0} = 1 = \binom{n}{n},$$

$$\binom{n}{k} = 0 \quad \text{for} \quad k > n \quad \text{and} \quad k < 0$$

$$\binom{n}{k-1} + \binom{n}{k} = \binom{n+1}{k}$$

$$\binom{-n}{k} = (-1)^{k}\binom{n+k-1}{k} \qquad \text{for} \quad n > 0$$

$$\sum_{k=0}^{n} \binom{n}{k} = 2^n, \quad \sum_{k=0}^{n} (-1)^k \binom{n}{k} = 0, \quad \sum_{k=0}^{n} \binom{n}{k}^2 = \binom{2n}{n}$$

$$\sum_{k=1}^{n} (-1)^{k-1} \frac{1}{k} \binom{n}{k} = 1 + \frac{1}{2} + \ldots + \frac{1}{n}$$

Stirling formula:

$$n! \sim (2\pi)^{\frac{1}{2}} n^{n+\frac{1}{2}} e^{-n}$$

$$\lim_{n \to \infty} \binom{n}{k} x^n = 0, \qquad |x| < 1$$

$$\lim_{n \to \infty} (1 + \frac{x}{n})^n = e^x, \qquad \lim_{x \to \infty} \frac{(1+x)^a - 1}{x} = a.$$

Special Functions:

Gamma functions:

$$\Gamma(z) = \int_0^\infty t^{z-1} e^{-t} \, dt \qquad (\text{Re } z > 0)$$

$$\Gamma(z+1) = z\Gamma(z)$$

For n positive integer: $\Gamma(n) = (n-1)!$, $\Gamma(1) = 1$,

$\Gamma(2) = 1$, $\Gamma(\frac{1}{2}) = \sqrt{\pi}$,

$$\Gamma(z+n) = (z+n-1) \ldots (z+1) z\Gamma(z)$$

$$\frac{d^2}{dz^2} \log \Gamma(z) = \sum_{n=0}^{\infty} \frac{1}{(z+n)^2} \; ;$$

Euler constant $\gamma = -\dfrac{d}{dz} \log \Gamma(z+1)\Big|_{z=0}$

$$= \lim_{n\to\infty} \left(1 + \frac{1}{2} + \ldots + \frac{1}{n} - \log n \right)$$

$$= 0.5772157\ldots .$$

Incomplete gamma function:

$$\Gamma_y(z) = \int_0^y t^{z-1} e^{-t}\, dt \qquad (\text{Re } z > 0)$$

$$\Gamma_y(z+1) = z\Gamma_y(z) - e^{-y} y^z \qquad (z \neq 0),$$

$$\frac{1}{(n-1)!}\, \Gamma_y(n) = e^{-y} \sum_{j=n}^{\infty} \frac{y^j}{j!} \qquad (n \text{ integer}).$$

Beta function:

$$B(p,q) = \int_0^1 x^{p-1} (1-x)^{q-1}\, dx, \qquad (\text{Re } p > 0, \quad \text{Re } q > 0)$$

$$B(p,q) = B(q,p), \qquad pB(p,q+1) = qB(p+1,q)$$

$$B(p,q) = \frac{\Gamma(p)\Gamma(q)}{\Gamma(p+q)} .$$

Incomplete beta function:

$$B_y(p,q) = \int_0^y x^{p-1}(1-x)^{q-1}\, dx$$

$$(\text{Re } p > 0, \quad \text{Re } q > 0, \quad 0 \le y \le 1)$$

Binomial Distribution

$$p(j) \;=\; \binom{n}{j} p^{j} (1-p)^{n-j}$$

n	j	0.10	0.25	0.40	0.50	p
1	0	.9000	.7500	.6000	.5000	
	1	.1000	.2500	.4000	.5000	
2	0	.8100	.5625	.3600	.2500	
	1	.1800	.3750	.4800	.5000	
	2	.0100	.0625	.1600	.2500	
3	0	.7290	.4219	.2160	.1250	
	1	.2430	.4219	.4320	.3750	
	2	.0270	.1406	.2880	.3750	
	3	.0010	.0156	.0640	.1250	
4	0	.6561	.3164	.1296	.0625	
	1	.2919	.4219	.3456	.2500	
	2	.0485	.2109	.3456	.3750	
	3	.0035	.0469	.1536	.2500	
	4	.0001	.0039	.0256	.0625	
5	0	.5905	.2373	.0778	.0312	
	1	.3280	.3955	.2592	.1562	
	2	.0729	.2637	.3456	.3125	
	3	.0081	.0879	.2304	.3125	
	4	.0004	.0146	.0768	.1562	
	5	.0000	.0010	.0102	.0312	

Poisson Distribution

$$p(j) = \frac{\mu^j}{j!} e^{-\mu}, \qquad\qquad F(j) = \sum_{i=0}^{j} p(i)$$

	$\mu = 0.1$		$\mu = 0.2$		$\mu = 0.3$		$\mu = 0.4$		$\mu = 0.5$	
j	p(j)	F(j)	p(j)	F(j)	p(j)	F(j)	p(j)	F(j)	p(j)	F(j)
	0.		0.		0.		0.		0.	
0	9048	0.9048	8187	0.8187	7408	0.7408	6703	0.6703	6065	0.6065
1	0905	0.9953	1637	0.9825	2222	0.9631	2681	0.9384	3033	0.9098
2	0045	0.9998	0164	0.9989	0333	0.9964	0536	0.9921	0758	0.9856
3	0002	1.0000	0011	0.9999	0033	0.9997	0072	0.9992	0126	0.9982
4	0000	1.0000	0001	1.0000	0003	1.0000	0007	0.9999	0016	0.9998
5							0001	1.0000	0002	1.0000

	$\mu = 0.6$		$\mu = 0.7$		$\mu = 0.8$		$\mu = 0.9$		$\mu = 1$	
j	p(j)	F(j)	p(j)	F(j)	p(j)	F(j)	p(j)	F(j)	p(j)	F(j)
	0.		0.		0.		0.		0.	
0	5488	0.5488	4966	0.4966	4493	0.4493	4066	0.4066	3679	0.3679
1	3293	0.8781	3476	0.8442	3595	0.8088	3659	0.7725	3679	0.7358
2	0988	0.9769	1217	0.9659	1438	0.9526	1647	0.9371	1839	0.9197
3	0198	0.9966	0284	0.9942	0383	0.9909	0494	0.9865	0613	0.9810
4	0030	0.9996	0050	0.9992	0077	0.9986	0111	0.9977	0153	0.9963
5	0004	1.0000	0007	0.9999	0012	0.9998	0020	0.9997	0031	0.9994
6			0001	1.0000	0002	1.0000	0003	1.0000	0005	0.9999
7									0001	1.0000

μ = 1.5		μ = 2		μ = 3		μ = 4		μ = 5	
p(j)	F(j)	p(j)	F(j)	p(j)	F(j)	p(j)	F(j)	p(j)	F(j)
0.		0.		0.		0.		0.	
2231	0.2231	1353	0.1353	0498	0.0498	0183	0.0183	0067	0.0067
3347	0.5578	2707	0.4060	1494	0.1991	0733	0.0916	0337	0.0404
2510	0.8088	2707	0.6767	2240	0.4232	1465	0.2381	0842	0.1247
1255	0.9344	1804	0.8571	2240	0.6472	1954	0.4335	1404	0.2650
0471	0.9814	0902	0.9473	1680	0.8153	1954	0.6288	1755	0.4405
0141	0.9955	0361	0.9834	1008	0.9161	1563	0.7851	1755	0.6160

μ = 1.5		μ = 2		μ = 3		μ = 4		μ = 5	
p(j)	F(j)	p(j)	F(j)	p(j)	F(j)	p(j)	F(j)	p(j)	F(j)
0.		0.		0.		0.		0.	
0035	0.9991	0120	0.9955	0504	0.9665	1042	0.8893	1462	0.7622
0008	0.9998	0034	0.9989	0216	0.9881	0595	0.9489	1044	0.8666
0001	1.0000	0009	0.9998	0081	0.9962	0298	0.9786	0653	0.9319
		0002	1.0000	0027	0.9989	0132	0.9919	0363	0.9682
				0008	0.9997	0053	0.9972	0181	0.9863
				0002	0.9999	0019	0.9991	0082	0.9945
				0001	1.0000	0006	0.9997	0034	0.9980
						0002	0.9999	0013	0.9993
						0001	1.0000	0005	0.9998
								0002	0.9999
								0000	1.0000

Exponential Function

x	e^x	e^{-x}	x	e^x	e^{-x}
0.00	1.0000	1.000000	1.00	2.7183	0.367879
0.05	1.0513	0.951229	1.05	2.8577	0.349938
0.10	1.1052	0.904837	1.10	3.0042	0.332871
0.15	1.1618	0.860708	1.15	3.1582	0.316637
0.20	1.2214	0.818731	1.20	3.3201	0.301194
0.25	1.2840	0.778801	1.25	3.4903	0.286505
0.30	1.3499	0.740818	1.30	3.6693	0.272532
0.35	1.4191	0.704688	1.35	3.8574	0.259240
0.40	1.4918	0.670320	1.40	4.0552	0.246597
0.45	1.5683	0.637628	1.45	4.2631	0.234570
0.50	1.6487	0.606531	1.50	4.4817	0.223130
0.55	1.7333	0.576950	1.55	4.7115	0.212248
0.60	1.8221	0.548812	1.60	4.9530	0.201897
0.65	1.9155	0.522046	1.65	5.2070	0.192050
0.70	2.0138	0.496585	1.70	5.4749	0.182631
0.75	2.1170	0.472367	1.75	5.7540	0.173774
0.80	2.2255	0.449329	1.80	6.0498	0.165299
0.85	2.3396	0.427415	1.85	6.3598	0.157237
0.90	2.4596	0.406570	1.90	6.6859	0.149569
0.95	2.5857	0.386741	1.95	.0237	0.142274
			2.00	7.3891	0.135335

The Normal Distribution

$$\phi(t) = \frac{1}{\sqrt{2\pi}} e^{-\frac{t^2}{2}}, \qquad \Phi(t) = \int_{-\infty}^{t} \phi(s)\, ds$$

t	$\phi(t)$	$\Phi(t)$	t	$\phi(t)$	$\Phi(t)$
0.0	0.398942	0.500000	2.3	0.028327	0.989276
0.1	.396952	.539828	2.4	.022395	.991802
0.2	.391043	.579260	2.5	.017528	.993790
0.3	.381388	.617911	2.6	.013583	.995339
0.4	.368270	.655422	2.7	.010421	.996533
0.5	.352065	.691462	2.8	.007915	.997445
0.6	.333225	.725747	2.9	.005953	.998134
0.7	.312254	.758036	3.0	.004432	.998650
0.8	.289692	.788145	3.1	.003267	.999032
0.9	.266085	.815940	3.2	.002384	.999313
1.0	.241971	.841345	3.3	.001723	.999517
1.1	.217852	.864334	3.4	.001232	.999663
1.2	.194186	.884930	3.5	.000873	.999767
1.3	.171369	.903200	3.6	.000612	.999841
1.4	.149727	.919243	3.7	.000425	.999892
1.5	.129518	.933193	3.8	.000292	.999928
1.6	.110921	.945201	3.9	.000199	.999952
1.7	.094049	.955435	4.0	.000134	.999968
1.8	.078950	.964070	4.1	.000089	.999979
1.9	.065616	.971283	4.2	.000059	.999987
2.0	.053991	.977250	4.3	.000039	.999991
2.1	.043984	.982136	4.4	.000025	.999995
2.2	.035475	.986097	4.5	.000016	.999997

Appendix C

Suggestions for Further Reading

There are really no elementary texts on stochastic processes. On the other hand, there is a great availability of books on statistics, and most of them contain chapters on probability and some of them contain a chapter or two on elementary stochastic processes (especially renewal and Markov chains).

Medium level books on stochastic processes:

1. N.T.J. Bailey, The Elements of Stochastic Processes with Applications to the Natural Sciences, J. Wiley, 1964.

2. E. Parzen, Stochastic Processes, Holden Day Inc., 1962.

3. S.M. Ross, Introduction to Probability Models, Academic Press, 1972.

4. See also articles in "Scientific American."

Advanced texts on stochastic processes:

5. A.T. Bharucha-Reid, Elements of the Theory of Markov Processes and their Applications, McGraw-Hill, 1960.

6. E. Cinlar, Introduction to Stochastic Processes, Prentice-Hall, Inc., 1975.

7. W. Feller, Introduction to Probability Theory and its Applications, Vol. 1, J. Wiley, 1957.

8. S. Karlin, H.M. Taylor, A First Course in Stochastic
    Processes, Academic Press, 1975.

9. N.U. Prabhu, Stochastic Processes, McMillan, 1965.

10. S.M. Ross, Applied Probability Models with Optimization
    Applications, Holden-Day, 1970.

There are also specialized works devoted to special
topics, like queueing, reliability, renewals as well as books
for mathematicians on Markov Processes, random walks, etc.
For a particularly successful blend of "ultra-advanced" theory
with significant practical applications, consult Volume 2 of
Feller.

Erlang distribution, 139, 155
ESP, 126
event, 3, 6
  complementary, 6
  sure, 5
  total, 5
events:
  combination of, 6, 15, 22, 35, 37, 183
  determined by random variables, 11, 12
  independent, 37
expectation, 21
  conditional, 200
  of compound random variable, 142-145, 158, 202
  of products, 40
  infinite, 53, 186, 244
  of sum, 40
explosion, 244
exponential distribution, 17, 21, 25, 242
exponential series, 136, 270
extinction, 256, 258
extreme life times, 42

factorial, 117
failure, 114
first crossing, 179, 184
  entrance, 218
  passage, 213, 218
  return, 218
forward equations, 251, 256
frequency, 8, 123
friends meeting, 63

gamma function, 272
Gaussian distribution, see normal distribution
  folded, 109
generating function (cost function), 151, 155-156, 230
geometric distribution, 129, 150-151, 180, 227